Milestones in Drug Therapy
MDT

St. John's Wort and its Active Principles in Depression and Anxiety

Edited by W. E. Müller

Birkhäuser Verlag
Basel · Boston · Berlin

Editor
Walter E. Müller
Pharmakologisches Institut für Naturwissenschaftler
Biozentrum Universität Frankfurt
Marie-Curie-Str.9
60439 Frankfurt
Germany

Library of Congress Cataloging-in-Publication Data
St. John's Wort and its active principles in depression and anxiety / W.E. Müller, editor.
p. ; cm.-- (Milestones in drug therapy)
Includes bibliographical references and index.
ISBN 3-7643-6160-3 (alk. paper)
1. Hypericum perforatum--Therapeutic use. 2. Affective disorders--Treatment.
I. Müller, Walter E. (Walter Erhard), 1947- II. Series.
[DNLM: 1. Hypericum. 2. Phytotherapy. 3. Anxiety Disorders--drug therapy. 4.
Depressive Disorder--drug therapy. QV 766 S143 2005]
RS165.H95S73 2005
615'.323624--dc22 20044062277

Bibliographic information published by Die Deutsche Bibliothek
Die Deutsche Bibliothek lists this publication in the Deutsche Nationalbibliografie;
detailed bibliographic data is available in the Internet at <http://dnb.ddb.de>.

ISBN 10: 3-7643-6160-3 Birkhäuser Verlag, Basel - Boston - Berlin
ISBN 13: 978-3-7643-6160-0

Printed on acid-free paper produced from chlorine-free pulp. TCF ∞

Printed in Germany
ISBN 10: 3-7643-6160-3
ISBN 13: 978-3-7643-6160-0
9 8 7 6 5 4 3 2 1 www. birkhauser.ch

Contents

List of contributors

Shyam S. Chatterjee, Pharmacology Laboratories, Research and Development Department, Dr. Willmar Schwabe GmbH&Co KG, Willmar-Schwabe-Straße 4, 76227 Karlsruhe, Germany; e-mail: shyam.chatterjee@schwabe.de
Michael Franklin, University of Oxford Department of Psychiatry, Warneford Hospital, Oxford OX3 7JX, United Kingdom; and School of Biological and Molecular Sciences, Oxford Brookes University, Oxford OX3 0BP, United Kingdom; e-mail: michael.franklin@psych.ox.ac.uk
Carla Gambarana, Department of Neuroscience, Pharmacology Unit, University of Siena, Via Moro 4, 53100 Siena, Italy; e-mail: gambarana@unisi.it
Daniela Giachetti, Department of Environmental Sciences, University of Siena, Via T. Pendola 62, 53100 Siena, Italy; e-mail: giachetti@unisi.it
Marco Gobbi, Istituto di Ricerche Farmacologiche "Mario Negri", Via Eritrea 62, 20157 Milano, Italy; e-mail: gobbi@marionegri.it
Kurt Hostettmann, Laboratoire de Pharmacognosie et Phytochimie, Section des Sciences Pharmaceutiques, Université de Genève, Quai Ernest Ansermet 30, CH-1211 Genève 4, Switzerland; e-mail: kurt.hostettmann@pharm.unige.ch
Andreas Johne, Institut für Klinische Pharmakologie, Universitätsklinikum Charité, Schumannstraße 20, 10098 Berlin, Germany; e-mail: andreas.johne@charite.de
Siegfried Kasper, Klinische Abteilung für Allgemeine Psychiatrie, Universitätsklinik für Psychiatrie, Medizinische Universität Wien, Währinger Gürtel 18–20, 1090 Wien, Austria; e-mail: sci-genpsy@meduniwien.ac.at
Jan Kehr, Department of Neuroscience, Karolinska Institutet, Retzius väg 8, 171 77 Stockholm, Sweden; e-mail: Jan.Kehr@neuro.ki.se
Tiziana Mennini, Istituto di Ricerche Farmacologiche "Mario Negri", Via Eritrea 62, 20157 Milano, Italy; e-mail: tiziana@marionegri.it
Walter E. Müller, Pharmakologisches Institut für Naturwissenschaftler, Biozentrum Universität Frankfurt, Marie-Curie-Str.9, 60439 Frankfurt, Germany; e-mail: PharmacolNat@em.uni-frankfurt.de
Michael Nöldner, Dr. Willmar Schwabe GmbH & Co KG, Willmar-Schwabe-Straße 4, 76227 Karlsruhe, Germany; e-mail: michael.noeldner@schwabe.de
Sven Ove Ögren, Department of Neuroscience, Karolinska Institutet, Retzius väg 8, 171 77 Stockholm, Sweden; e-mail: Sven.Ove.Ogren@neuro.ki.se
Manfred Schubert-Zsilavecz, Institute of Pharmaceutical Chemistry, Johann Wolfgang Goethe University Frankfurt/Main, Marie-Curie-Str. 9, 60439

Frankfurt/Main, Germany; e-mail: schubert-zsilavecz@pharmchem.uni-frankfurt.de

Volker Schulz, Oranienburger Chaussee 25, 13465 Berlin, Germany; e-mail: V.Schulz.Berlin@t-online.de

Kristina Treiber, Pharmakologisches Institut für Naturwissenschaftler, Biozentrum Universität Frankfurt, Marie-Curie-Str.9, 60439 Frankfurt, Germany; e-mail: K.Treiber@em.uni-frankfurt.de

Hans-Peter Volz, Krankenhaus für Psychiatrie und Psychotherapie Schloss Werneck, Balthasar-Neumann-Platz 1, 97440 Werneck, Germany; e-mail: hans-peter.volz@kh-schloss-werneck.de

Jean-Luc Wolfender, Laboratoire de Pharmacognosie et Phytochimie, Section des Sciences Pharmaceutiques, Université de Genève, Quai Ernest Ansermet 30, CH-1211 Genève 4, Switzerland; e-mail: jean-luc.wolfender@pharm.unige.ch

Mario Wurglics, Institute of Pharmaceutical Chemistry, Johann Wolfgang Goethe University Frankfurt/Main, Marie-Curie-Str. 9, 60439 Frankfurt/Main, Germany; e-mail: Wurglics@pharmchem.uni-frankfurt.de

Takashi Yoshitake, Department of Neuroscience, Karolinska Institutet, Retzius väg 8, 171 77 Stockholm, Sweden; e-mail: Takashi.Yoshitake@neuro.ki.se

St. John's Wort and its Active Principles in Depression and Anxiety
Edited by W. E. Müller

Introduction and historical overview

Walter E. Müller

Pharmakologisches Institut für Naturwissenschaftler,Biozentrum Universität Frankfurt, Marie-Curie-Str. 9, 60439 Frankfurt, Germany

The therapeutic properties and advantages of the medicinal herb St. John's Wort, and preparations derived from it, have been known for more than 2,000 years. However, the specific use of the plant as an antidepressant drug, or as a herbal preparation to treat depressive illness, is relatively recent and has come to the fore only in the last 20 years. St. John's Wort was first mentioned in the Roman times by Pliny the Elder in the 1st century AD [1]; interestingly it was not to treat melancholia, although the condition was known at that time, but as a "seed of bracing quality which checks diarrhoea and promotes urine and is taken with wine for bladder troubles". Around the same time, Diascorides, a Roman army doctor born in Greece recommended the imbibition of St. John's Wort with special liquids "to expel many choleric excrements" [1]. Over 1,000 years later, the famous German physician Paracelsus was one of the first to mention St. John's Wort as a remedy to treat mental disturbances and also other diseases. He specifically recommended it for three conditions, "wounds, parasites, and phantasmata", where the latter meant psychoses such as hallucinations and delusion rather than depression [1]. He also mentioned melancholia, but did not specifically suggest St. John's Wort as treatment. Inspired by Paracelsus, Angelo Sala was probably the first practitioner to specifically recommend St. John's Wort for depression or melancholia as it was known at that time. In 1630 he wrote (cited from [1]):

> "St. John's Wort has a curious, excellent reputation for the treatment of illnesses of the imagination, which are known by some as phantasmata and by others as mad spirits, and for the treatment of melancholia, anxiety and disturbances of understanding, which sometimes affect highly intelligent people whose primary personality is not melancholic and in whom you do not see persistent melancholic humour. St. John's Wort cures these disorders as quick as lightning. It takes a day and a night. With the same power it works against the symptoms caused by witches in a way that is superior – as best I can tell – to the effects of any other type of plant or medication, though these may be very highly respected."

Most importantly, especially considering the present discussion about the relevant active constituents of St. John's Wort, he suggested that it should best be given as a tincture made from the fresh petals and leaves by using brandy. He also recommended making the tincture by warming it and to cover the preparation vessel to protect it from daylight. We know today that all these conditions are relevant and that the constituents such as hyperforin and adhyperforin that are not very water soluble are present in goodly amounts. About 200 years later, the German poet-physician J. Kerner also mentioned St. John's Wort in his writings for the treatment of mood disorders (melancholia). It is quite interesting that for a further century, St. John's Wort did not receive much more attention, and it was not used as an antidepressant when modern psychiatry first introduced drug-treatment around 1900; at this time drugs such as opium, barbiturates, and other sedatives gained a reputation as antidepressant treatments. Probably the last entry into older history of St. John's Wort came from the German physician K. Daniel, who not only carried out animal experiments but also described in fine detail his experience with St. John's Wort extract in the treatment of about 20 depressed patients. Unfortunately, his rather timely publication, which came out just prior to World War II, did not receive much attention in the years following (Daniel, 1939) [2].

The modern history of St. John's Wort extract probably began in 1984, when Commission E of the former German Federal Health agency, which was determining recommendations for the use of herbal drugs, published a positive monograph about St. John's Wort, recommending it use for psychoautonomic disturbances, depressed mood, nervousness, and anxiety. According to current standards, this recommendation was based mainly on traditional use and experience, and not on sound scientific data or even clinical studies. However, initiated by this monograph (although this is just an assumption), several smaller companies started to carry out controlled clinical studies (against placebo or against an active comparator, at that time usually a tricyclic antidepressant) in depressed patients. Although usually too small and statistically underpowered according to current standards, these studies gave the first evidence for the clinical usefulness of St. John`s Wort extract in the treatment of depression in Germany, using brands with extract doses of at least a few hundred milligrams per day. These data finally led to the introduction of standardised St. John's Wort extract brands which allowed a daily dosage of 600–900 mg. Using these new preparations, new controlled clinical studies were performed, which confirmed the clinical efficacy of the herbal drug in mild-to-moderate depressive illness. A first overview about the clinical data and some preliminary pharmacological findings was published in English as a supplement to the *Journal of Geriatric Psychiatry and Neurology* in October 1994 [3]. These data finally led to an overwhelming acceptance of St. John's Wort preparations in Germany as an alternative to synthetic antidepressants. In the late 1990s, St. John's Wort preparations made up as much as 25% of all prescriptions filled out by German doctors for antidepressants. The use and popularity of St. John's Wort as an antidepressant has also increased in many other countries all over the world [1, 4, 5], but no other country has posted such a high level of acceptance as Germany. This is probably because in most other countries St. John's Wort extract preparations are sold as food supplements with little regulatory and quality control, in contrast to Germany and a some other countries where herbal medicines are sold as OTC drugs (which could still be reimbursed by the health insurance system) with excellent quality standards. The increasing popularity of this herbal drug has led to much more detailed clinical, biochemical, pharmacological, and pharmacokinetic studies. The various studies have led to the publication of three supplements of *Pharmacopsychiatry*, a German-based international journal for clinical psychopharmacology and biological and clinical psychiatry [6–8] and to a rapidly increasing number of other scientific publications about all aspects of St. John's Wort extract.

Taken together, the modern history of St. John's Wort extract is a success story The reasons behind the success are not only rationale, e.g., its efficacy and its relatively low side effect profile (even if we know today that it is not free of side effects and possible drug interactions), but are also to some part emotional. Antidepressants are not liked by patients, not only because of side effects and because of concerns about drug dependency (which is definitively not the case) but also because by taking the antidepressant, the patient in some ways needs to accept the disease and the diagnosis of depression. This is not the same for the herbal antidepressant. Accordingly, their acceptance by patients is often much better than for the "bad chemical" antidepressant drugs. A typical case study could be somatoform disorders, where two recently published positive clinical studies with St. John's Wort have shown that although these patients are traditionally very difficult to convince to take psychotropic drugs, they will be more likely to accept the herbal drug. Patients will be much more likely to accept the natural compound [5], even if, as we will learn later, the herbal preparation acts on the same neurochemical pathways in the brain as the synthetic antidepressants. Thus, many patients who can not be reached by synthetic antidepressants can be reached by St. John's Wort as initial treatment and, even more importantly, can sometimes be continued on other antidepressants if St. John's Wort is not sufficiently efficacious. Thus, although it is known that the use of St. John's Wort can produce some side effects and drug interactions, its important role in the initial therapeutic strategy in depression and related mood disorders lies in its positive acceptance by patients. It is not so much an alternative for the more than 20 chemical antidepressants available, but it is a further option especially for patients who are rather difficult to convince to accept other treatments. With less than 50% of depressed patients treated adequately, as is the case for most western societies, it is believed that there is much scope for such a "natural" treatment, if it is effective and its risk/benefit ratio is acceptable.

Several years have passed since the last comprehensive review of St. John's Wort was published [8]. I gratefully accepted the offer by the editors of the respective series *Milestones in Drug Therapy* to edit a new issue of this series devoted to all aspects of St. John's Wort from phytochemistry, pharmacology, clinical efficacy, site effects, and drug interaction which will give the interested reader the most recent overview about all aspects of this herbal antidepressant. As editor I would like to thank all authors for their interesting and learned contributions, the series editors for the opportunity, and the publisher for their cooperation and the speedy realization of the project.

Walter E. Müller, Frankfurt
December 2004

References

1 Rosenthal N (1995) *St. John's Wort. The herbal way to feeling good.* Harper Collins Publishers, New York
2 Daniel K (1939) Johanniskraut bei psychischen Störungen. *Hippocrates* 10: 929–932
3 Jenike MA (1994) *Hypericum*: A novel antidepressant. *Journal Geriatric Psychiatry and Neurology* 7 (Suppl 1): 1–68
4 Rosenthal N (1998) *St. John's Wort. Your natural Prozac.* Thorsons, London
5 Cass H (1998) St. John's Wort. *Nature's blues buster.* Avery Publishing Group, New York
6 Müller WE, Kasper S (eds) (1997) *Hypericum* extract (LI 160) as a herbal antidepressant. *Pharmacopsychiatry* 30 (Suppl II): 72–134
7 Müller WE, Chatterjee SS (eds) (1998) Hyperforin and the antidepressant activity of St. John's Wort. *Pharmacopsychiatry* 31: 1–60
8 Müller WE, Chatterjee (eds) (2001) Biocenter symposium on drug therapy. Advances of St. John's Wort research. *Pharmacopsychiatry* 34 (Suppl I): 2–156

Table 1. The history of the use of St. John's Wort as an antidepressant. A more detailed account is given by Rosenthal [1, 4]

Historical outline of the medical use of St. John's Wort as antidepressant	
Paracelsus (1491–1541)	St. John's Wort is recommended for three different conditions: wounds, parasites, and "phantasmata". With the latter condition he rather refers to psychotic than depressed states. Nevertheless, he also uses the terms "healing of the soul" and "arnica for the nerves".
Angelo Sala (1630)	He also recommends St. John's Wort as treatment for phantasmata and specifically for melancholia. He proposes it should be best given as an extract made with brandy.
J. Kerner (1786–1862)	The German poet-physician J. Kerner reported on the use of St. John's Wort in the treatment of mood disorder (melancholia).
K. Daniel (1939)	First description of a modern treatment of depressed patients with a St. John's Wort extract.
BGA[1] (1984)	Positive monograph of the St. John's Wort for the following uses: psychoautonomic disturbances, depressed mood, nervousness, anxiety.
Several German authors (1985–1991)	First clinical studies in depressed patients with standardised high-dose St. John's Wort extracts in Germany.

[1] Bundesgesundheitsamt, the former German Federal Health Agency (Bundesanzeiger, 1984)

Phytochemistry

Kurt Hostettmann and Jean-Luc Wolfender

Laboratoire de Pharmacognosie et Phytochimie, Section des Sciences Pharmaceutiques, Université de Genève, Quai Ernest Ansermet 30, CH-1211 Genève 4, Switzerland

Introduction

St. John's Wort (*Hypericum perforatum*) is a member of the Hypericaceae family. The genus *Hypericum* contains around 400 species divided in 30 subgroups that are spread throughout temperate and tropical areas worldwide. More than 70 species have been investigated from a phytochemical viewpoint. *H. perforatum* is one of the most widely distributed species along with *H. crispum* and *H. hirsutum*. This plant, as other members of the Hypericaceae, is a prolific producer of secondary metabolites [1] and has been the topic of numerous phytochemical investigations due its ancient use as a vulnerary and especially, more recently, its widespread application as an antidepressant phytopreparation. Today, the registered phytomedicines consist of an alcoholic extract of the dried flowering tops of the plant. At least height natural product chemical classes are present in this plant; they include naphthodianthrones, phloroglucinols, flavonoids, biflavonoids, xanthones, proanthocyanidins, acid phenols as well as essential oils. Several review have already dealt with the constituents [2–4].

In spite of all the chemical studies performed on this widespread medicinal plant the major limitation to a rational exploitation of this plant is our still-limited knowledge on its active constituents. In this Chapter, a review of the current knowledge of the chemical composition of St. John's Wort is given, as well as different aspects related to the standardisation of the extracts.

Naphthodianthrones

One of the most characteristic constituents of *H. perforatum* is the naphthodianthrone hypericin. This red pigment, which is exuded when the buds and flowers are squeezed, was associated with the blood of St. John the Baptist. *H. perforatum* pigments have been known for centuries and were considered to be responsible for the colouration of *H. perforatum* oil used as a popular healing remedy in the Middle Ages as anti-inflammatory and vulnerary for topical applications.

The first scientific investigation performed on the red pigments from *Hypericum* species is mentioned in 1830 by Buchner who gives the name ‘Hypericumrot’ to the pigment [5]. In 1895 already, the absorption spectrum (alcoholic solution) was investigated. The name ‘hypericin’ was given by Cerny in 1911, who was able to analyse its molecular formula ($C_{16}H_{10}O_5$). At this time 1.2 g were isolated from 1 kg of dried flowers [4]. After initially being wrongly attributed to the anthocyanidin class of compounds in 1927 [6], the structure of hypericin (Fig. 1a) was established in 1953 as 10-11-dimethyl-1,3,4,6,8,12,-hexahydroxynaphthodianthrone by Brockman et al. [7], who also described its total synthesis. Since this discovery, different hypericin derivatives were described in *Hypericum* species between 1957 and 1976 [7–14]. A survey of more than 200 *Hypericum* species demonstrated that these pigments are not distributed in all species and, apart from a few scattered exceptions, practically all hypericin-containing species belong to the sections *Euhypericum* and *Campylosporus* of Keller’s classification [15]; in *H. perforatum*, this is mainly hypericin (Fig. 1a) and pseudohypericin (Fig. 1b). Protophypericin (Fig. 1c) and protopseudohypericin (Fig. 1d) are converted into hypericin and pseudohypericin when the extracts are exposed to light [16]. More recently, an x-ray diffraction of the pyridinium salt of hypericin was described showing that the molecule is distorted and has a helical twist [17]. The solubility of hypericin strongly depends on its form; the free hypericin is only slightly soluble in polar solvents, while it forms a salt with inorganic bases (pH 4–11) which are generally much more soluble [18]. Hypericin, which has been used in most of the biological and clinical trials, is in fact a

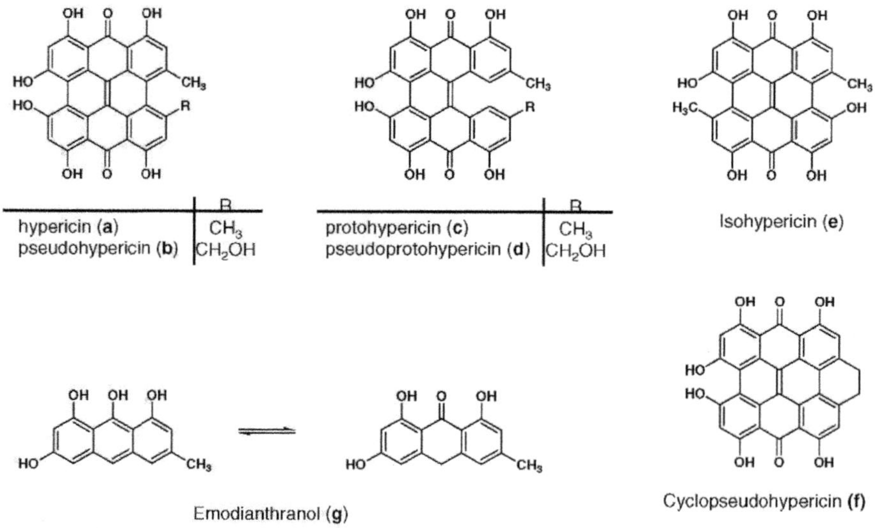

Figure 1. Naphthodianthrones.

monosodium salt. In the plants, it occurs mainly as a potassium salt [18]. These salts in inorganic solvent are red, highly absorbent (in EtOH, $\lambda max = 545$ and 590 nm $\varepsilon = 52{,}000$) and exhibit red fluorescence [17].

Biological activities of hypericin

At first, hypericin was wrongly attributed as being responsible for the antidepressant activity of St. John's Wort extracts. It was claimed that this compound exhibited MAO-A inhibiting properties [19], but finally it was demonstrated that this was due to an impurity of the hypericin that was used for the assay [20]. It has since been established that hypericin seems not to play a major role in the antidepressant effect of St. John's Wort – this pigment is, however, responsible for various other interesting biological effects that include photodynamic properties, antiviral and potential antineoplastic activities [18].

Hypericin and its salts are photodynamically active. These photosensitising properties were first observed in cattle with white or light coloured coats which feed on *Hypericum* and which developed a disease called hypericism. The *in vivo* effects caused skin erythema and oedema. In cattle, the following symptoms were observed: skin efflorescence in the form of blisters, psychomotor excitement and in grave cases, hemolysis, epileptic fits and death. Hypericism is derived, to a large extent, from the generation of singlet oxygen [21] by hypericin or its derivatives upon irradiation to visible light. This effect has been confirmed also *in vitro* where it could be demonstrated that hypericin oxidised tryptophan or fatty acids by a profile that implies singlet oxygen (type II mechanism) [21]. It is important to note that the photosensitisation effects only occur when the plant is ingested. In the case of consumption of phytopreparations containing *H. perforatum*, a risk of photosensitisation in human exists but it is limited. It is recommended not to be exposed intensively to sunlight upon treatment with St. John's Wort. The topical application of *Hypericum* oil for wound healing is, however, perfectly safe.

Hypericin appears also to be an effective virucidal agent, which directly inactivates a broad range of viruses [22–25] and retroviruses [26–30]. These include the murine Friend [30] and Rausher viruses [25], equine infection anaemia viruses [26], murine immunodeficiency virus [27], murine [23] human [22] cytomegaloviruses, influenza [25], vesiculostomatitis [28] virus, sendai [28], herpes [25] and ducks hepatitis B viruses [24]. In many cases, the photodynamic action in the virucidal activity of hypericin was noted. Its mode of action can be a direct action on the virus, possibly on the membrane constituents, but it can also be directed at virus-infected cells [18].

This important antiviral action has resulted in a great deal of interest for hypericin as a potential compound for curing AIDS [31]. In this respect however, a first clinical trial has revealed that following the treatment of 30 patients, significant phototoxicity was observed but no antiretroviral activity in the limited number of patients studied was recorded [32].

Besides these antiviral activities hypericin exhibits an interesting potential as antineoplastic agent. It has been shown that hypericin inhibits succinoxidase, an enzyme that is suggested to be positively related to neoplastic changes. The lytic phase of the cytotoxicity reaction of CD8 lymphocytes is also inhibited by hypericin, which is significant for the cellular immune system. Protein kinase and other kinases have also shown to be inhibited by hypericin [18].

The only adverse effect reported for hypericin is hypericism. Photosensitisation was studied in humans after the intake of pure synthetic hypericin. This phenomenon was found to be transient and diminishes a few days after hypericin is discontinued.

Until recently, hypericin was considered the undiscussed active ingredient of *Hypericum* species and used for the standardisation of phytopreparations [3]. The 'total hypericins' (hypericin + pseudohypericin) is around 0.1–0.15% in the extracts [33]. Today, its role in antidepressant activity is much less clear, but it remains an interesting marker for the standardisation of the extracts since it could be easily detected by UV (see standardisation section below).

Minor hypericin derivatives such as isohypericin (Fig. 1e) and cyclopseudohypericin (Fig. 1f) have also been reported in *H. perforatum* [2]. Besides these naphthodianthrones, anthranolic constituents, namely emodin anthranol (Fig. 1 g) have also been characterised in *H. perforatum.* It is not clear however if this anthranol derivative is the precursor of hypericin in the plant [4].

Phloroglucinols

H. perforatum mainly contains two acylated phloroglucinol derivatives hyperforin (Fig. 2a) and adhyperforin (Fig. 2b). Hyperforin is known for its remarkable antibacterial properties [34] and recent studies indicated that it might be an essential component for the antidepressive activity, as this compound represents the major reuptake inhibiting constituent of the extract of St. John's Wort [35]. Hyperforin was studied by Russian researchers between 1971–1975 [36–39]. The absolute configuration of hyperforin has been established by x-ray analysis [40]. This compound is chemically related to the hop bitter principle humulon and lupulon and has only been found in *H. perforatum.* The structure of adhyperforin, which is less abundant than hyperforin in the extracts, was described later [41].

The chemical instability of hyperforin has excluded for a long time its definitive pharmacological evaluation. A successful method for its isolation purification and storing has only recently been developed [1]. The procedure involves a separation of the CO_2 extract, which is enriched in phloroglucinol derivatives, by high speed countercurrent chromatography followed by immediate dilution in MeOH and storing at –20 °C. For easy handling, hyperforin was derivatised as organic or inorganic salts or isolated by countercurrent chromatography of its dicyclohexylammonium salt just before its use for biological evaluation [1].

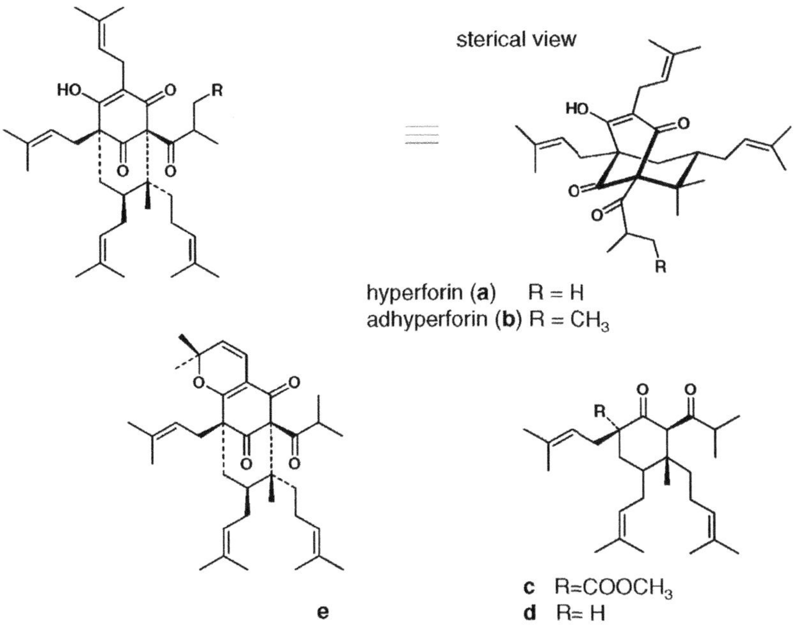

Figure 2. Phloroglucinols.

The chemistry of the acylated phloroglucinols in *H. perforatum* is complex. A liquid chromatography/thermospray mass spectrometry (LC/TSP-MS) analysis of a CO_2 extract of *H. perforatum* flowering tops revealed the presence of at least 14 phloroglucinol-type compounds, the most abundant being hyperforin and its homologue adhyperforin (10%). The other constituents are not well defined and appeared to be constituents showing the loss of an isoprenylic chain or showing 16 mass units more than hyperforin based on the on-line MS data recorded [1].

Hyperforin is found to be present at 2.0–4.5% in the extracts while adhyperforin represent 0.2–1.9% [2]. Some minor hyperforin (Fig. 2c–e) analogues have been reported [42]. Compounds 2c and 2d can be considered as intermediates in the metabolic pathway, while 2e was also found among the products obtained after chemical oxidation of hyperforin (Fig. 2).

Hyperforin instability

As mentioned above, hyperforin is unstable and its lability to oxidative degradation poses problems for standardisation and may also dramatically affect the pharmacological activity of the extracts. The identity of the major oxidised forms of hyperforin (Fig. 3a–j) has been recently reported, but it is not clear if these compounds are genuine plant constituents or artefacts formed *en route* to

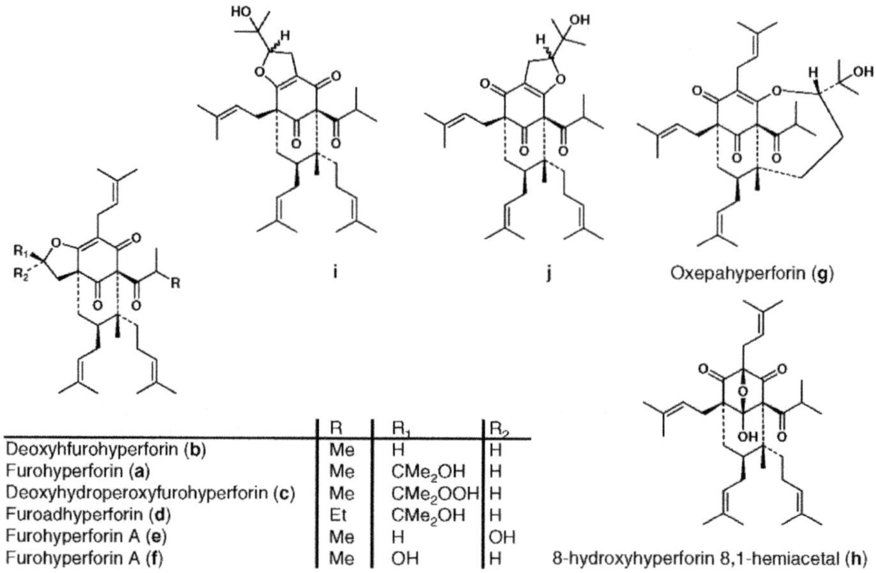

	R	R_1	R_2
Deoxyhfurohyperforin (**b**)	Me	H	H
Furohyperforin (**a**)	Me	CMe_2OH	H
Deoxyhydroperoxyfurohyperforin (**c**)	Me	CMe_2OOH	H
Furoadhyperforin (**d**)	Et	CMe_2OH	H
Furohyperforin A (**e**)	Me	H	OH
Furohyperforin A (**f**)	Me	OH	H

Figure 3. Oxidised forms of hyperforin and derivatives.

the degradation of the natural product [43–48]. Hyperforin is stable in protic solvents and unstable in apolar solvents such as *n*-hexane [43].

The broad shape of most of the ^{1}H-NMR signals, and the poor resolution of ^{13}C-NMR denoted that hyperforin is a mixture of tautomers. On the contrary, the oxidised forms show sharp signals that are indicative of a covalent block of the tautomeric equilibrium [1]. Different studies for a better understanding of the relationship between hyperforin and its oxidised forms have been conducted. The oxidation of pure hyperforin was performed by dissolving it in a hexane solution overnight at room temperature. The resulting mixture was analysed by LC/electrospray ionisation mass spectrometry (ESI-MS) [43] and LC/nuclear magnetic resonance (NMR) [48]. The results showed that the oxidised form obtained in this way (Fig. 3i and j) were isomers of hydroxyfurohyperforin (Fig. 3a) with dihydrofuran ring closure in two different positions, which can be explained by the keto-enol tautomerism of hyperforin [1].

In spite of the very important number of publications related to the biological activities of hyperforin, the chemical modification of this acylated phloroglucinol has so far received little attention. Furthermore, natural analogues are difficult to study since St. John's Wort has yielded them in a limited number and only at low abundance [1]. Recent structure activity relationship studies by Verotta [49] on different oxidised analogues of hyperforin have demonstrated that the covalent block of the highly reactive α-substituted enolised β-dicarbonyl system as an ether group is detrimental to the biological activity (inhibition of the synaptosomal accumulation of serotonin). Indeed, all natural and synthetic oxygenated hyperforin derivatives showed a lower activity on the reuptake of serotonin. This suggests a probable specific role of the enolised β-diketone moiety of hyperforin in the activity.

Whatever the origin, the oxidised products, their structural characterisation and availability have an obvious relevance for the standardisation of St. John's Wort extracts and the study of hyperforin metabolism.

Flavonoids

The main flavonoid present in *H. perforatum* is quercetin (Fig. 4a) and its related glycosides (2–4%). Quercetin-3-O-galactoside (Fig. 4e) was the first one to be described in *H. perforatum* and it was wrongly named hypericin in the original paper, which reported its isolation [50]. This compound was re-isolated later by Sprecher and correctly named hyperoside [51]. Other widespread quercetin glycosides (quercetin-3-O-rhamnoside (quercitrin [Fig. 4b]), quercetin-3-O-glucoside (isoquercitirin [Fig. 4c]) and quercetin-3-O-rutinoside (rutin [Fig. 4d]) were also reported in this plant [3]. Minor flavonoid aglycones were also reported in *Hypericum* species, these include dihydroquercetin, luteolin, kaempferol and myricetin [4]. Concentration of the flavonoids rutin (1.6%), hyperoside (0.9%), and isoquercitirin (0.3%) have been reported [52].

Flavonols

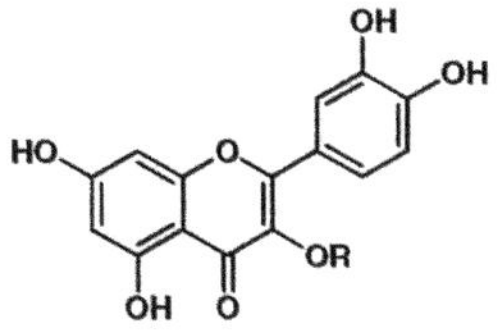

	R
Quercetin (**a**)	H
Quercitrin (**b**)	Rha
Isoquercitirin (**c**)	Glc
Rutin (**d**)	Rha-Glc
Hyperoside (**e**)	Gal

Xanthones

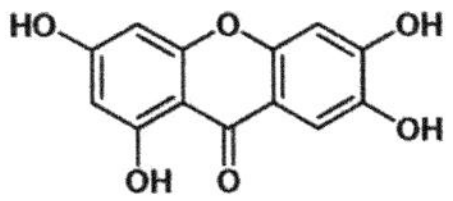

1,3,6,7-tetrahydroxyxanthone (**h**)

Biflavonoids

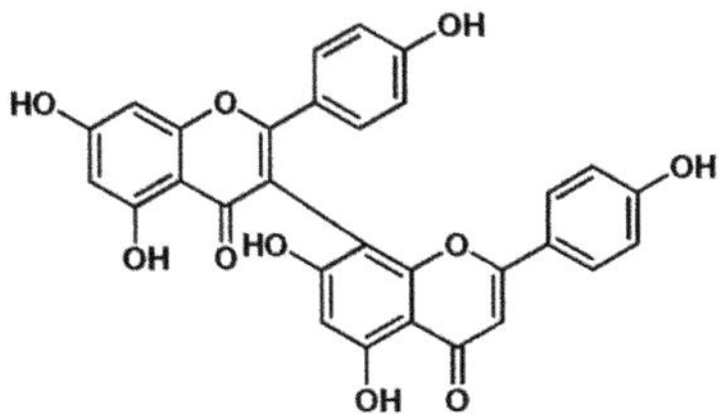

I3,II8-biapigenin (**f**)

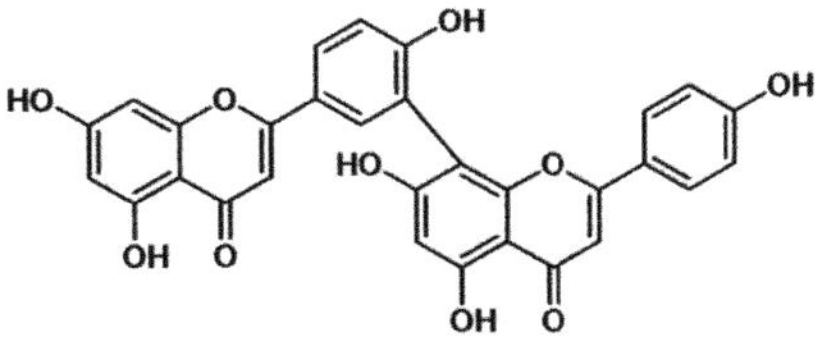

I3',II8-biapigenin (**g**)

Figure 4. Flavonoids and related polyphenols.

More recently, dimeric flavonoids I 3,II 8 biapigenin (Fig. 4f) (0.1–0.5%) [53] and I 3',II 8-biapigenin(amentoflavone) (Fig. 4 g) (0.01–0.05%) [54] have been isolated in small amounts.

The flavonoid fraction of *H. perforatum* displayed interesting *in vitro* inhibition of MAO-A [55]. It was established that these compounds were present at a 20% level at least in the fraction of hypericin that was previously tested by Suzuki [19], which wrongly attributed MAO-A activity to hypericin. However, according to the work of Bladt *in vivo* and *ex vivo*, even if the flavonoid fraction was found to be the most active, the antidepressant activity could not be expressed in terms of MAO inhibition only [20].

Proanthocyanidins

Proanthocyanidins represent approximately 12% of dried weight of aerial biomass including the seeds [56]. The proanthocyanidin mixture contains only catechin (Fig. 5a) or epicatechin (Fig. 5b) derivatives (dimers, trimers, tetramers and high polymers) since the acid hydrolysis resulted only in cyanidin [57]. This class of flavonoids were endowed with vascular activity resembling that of *Crataegus* species [3].

Xanthones

A 1,3,6,7-tetrahydroxyxanthone (Fig. 4h) was found in trace amounts in the extract of *H. perforatum* [58]. This compound, as other xanthones mainly occurring in the Gentianaceae as well as other Guttiferae species, was found to exhibit interesting IMAO properties [59], but was present in too small quantities to justify a possible role in the activity of St. John's Wort [60].

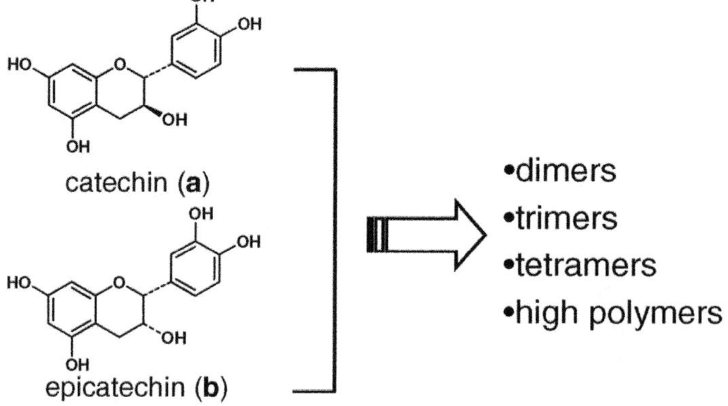

Figure 5. Proanthocyanidins.

Acid phenols

Various widespread acid phenols such as p-coumaric (Fig. 6e), ferulic (Fig. 6a), isoferulic (Fig. 6b), caffeic (Fig. 6c) and chlorogenic acids (Fig. 6d) have been described in several occasions in *H. perforatum* [61].

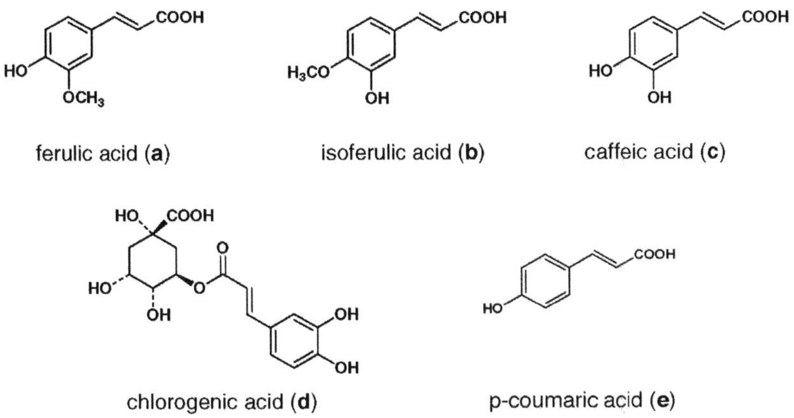

Figure 6. Acid phenols.

Essential oils

Besides the early work performed on hypericin by Buchner [5], in 1904 the phytochemical investigation of *H. perforatum* began with the determination of the content of essential oils. The species name 'perforatum' relates to the presence under the leaves of clear dots which allows light to pass through. These punctuations are not holes in the leaves but oily inclusions that contain the essential oils. A needle perforation of these inclusions allows the collection of the clear liquid that is the essential oil [4].

The content of the essential oils of St. John's Wort varies from 0.1–0.35% depending on the harvesting period and the quality of aerial part. The major constituent was found to be α-pinene (Fig. 7a) [3]. In the 1960s Ourisson and Mathis performed several gas chromatography (GC) studies on the essential oils of different *Hypericum* species [62–64]; in their study the main compounds identified were 2-methyloctane (Fig. 7b), n-nonan, n-undecan, α-pinene, β-pinene (Fig. 7c), limonene, myrcene caryophyllene, α-terpineol (Fig. 7d), geraniol, octanal, decanal and 2-methyldecane; the major component being 2-methyloctane (>30%). Recently, the composition of the volatile oils of the aerial part of St. John's Wort collected in six localities in the south of France were compared by gas chromatography/mass spectrometry (GC-MS) [65]. This latter study identified between 29 and 41 components. The

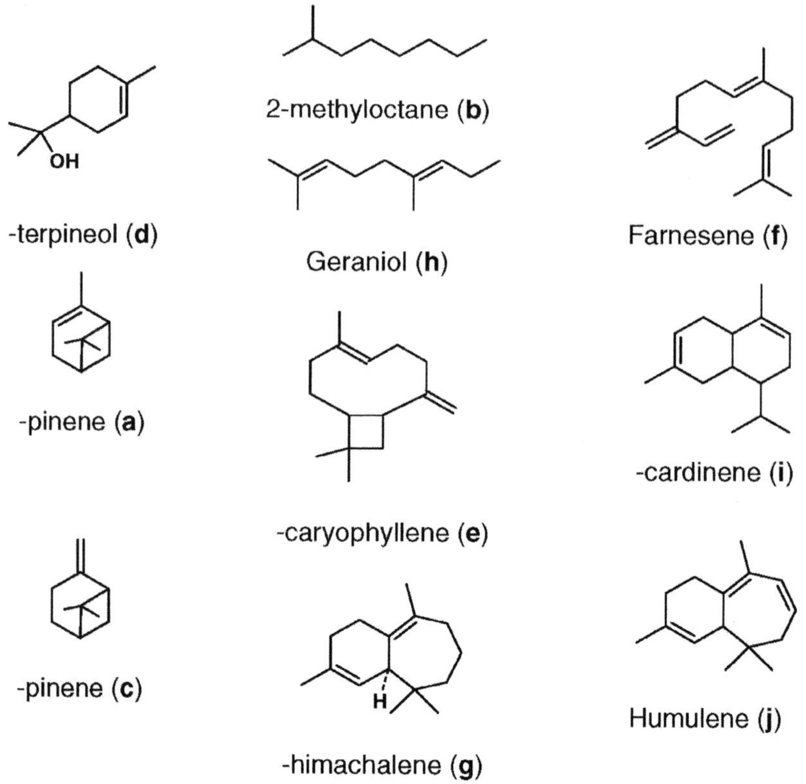

Figure 7. Essential oil constituents.

main constituents were found to be oxygenated and non-oxygenated sesquiterpenes. This is in contradiction with previous studies that revealed mainly the presence of monoterpenes. 14 compounds were present in the essential oils of each tested population, namely β-caryophyllene (Fig. 7e), caryophyllene oxide, (E)-β-farnesene (Fig. 7f), γ-cadinene, δ-cadinene, *ar*-curcumene, *cis*-calamenene, branched tetradecanol, spathulenol, nerolidol, α-cadinol, 2-methyldodecane and dodecanol; however, these compounds were present at different rates. Germacrene D and bicyclogermacrene were present in important quantities in some samples but were absent in others. It is interesting to notice the presence of α- or β-himachalene (Fig. 7 g), rare in plant chemistry, in all the *H. perforatum* var. *perforatum* oils but not in the *angustifolium* variety. The oil of the variety *angustifolium* is poor in farnesene forms, unlike the other oils, which are rich in (E)--, (E,E)-α- or (Z,E)-α-farnesene. In the oil of this variety, there is little content of β-caryophyllene and caryophyllene oxide [65]. Hence, a chemical difference between the two varieties may be revealed as demonstrated between *H. perforatum* var. *perforatum* and var. *angustifolium* from Serbia [66].

Other lipophilic constituents

In the lipophilic extract, apart from the compounds described above, several alkanes [67] and long chain fatty alcohols are present [68]. The study of the n-alkanes revealed the presence of all members in the series C_{16} C_{19} and the prevailing n-alkane was found to be nonacosane. Fatty acids like lauric, palmitic acids and some carotenoids (lutein, violaxanthin, cis throllixanthin, throllichromone) have been also identified [69].

St. John's Wort preparations

St. John's Wort oil

St. John's Wort oil is a crude product extracted from *H. perforatum* with vegetable oil. It has been known as a popular healing remedy since the Middle Ages and is mainly used externally for the treatment of wounds, especially burn wounds, bruises and swellings [70]. This preparation is endowed with an exceptionally high antibacterial activity. St. John's Wort oil is prepared by maceration of fresh flowers of *H. perforatum*, collected when the seeds begin to mature, in sunlight for several weeks. It acquires a brilliant red colour and orange–red fluorescence. The *Pharmacopoeia Wirtenbergica* (1847 edition) directs the extraction of dried flowers of St. John's Wort by digesting with hot olive oil for 3 h. The oil obtained is yellow due to the colour of the flower pigments.

The red colour and fluorescence of St. John's Wort oil was generally ascribed to hypericin and the pigments are determined by simple direct UV absorption measurements at 590 nm and expressed as hypericin. With specific methods, however hypericin is not measurable in the St. John's Wort oil. Only lipophilic compounds with a hypericin-like colour and fluorescence are present. These compounds were originally considered to be lipophilic substituted hypericin that were also present in St. John's Wort flowers, but it has now been established that I 3', II 8-biapigenin and 1,3,6,7-tetrahydroxy-xanthone are responsible for the colour and fluorescence [71].

Hyperforin could not be identified in the oil because of its high lipophilicity, which hampers its isolation from the oily matrix, and its low stability. It is probable that degradation products of hyperforin endowed with potent antibacterial activity are present [3].

Hypericum perforatum *phytopreparations*

The extracts used in therapy today are prepared by extracting the upper aerial parts of *H. perforatum,* collected just before or during blossom, with mixtures of ethanol/water or methanol:water [72]. The content of the extracts is directly related to the harvesting period, drying process and storage. Table 1 [61]

Table 1. Variation of the composition of flowers in 50 specimens [61]

	µg/flowers	
constituents	max	min
hypericin	23	3
pseudohypericin	64	11
hyperforin	607	206
13',II8-biapigenin	71	11
rutin	61	19
hyperoside	140	44
isoquercitrin	107	15
quercitrin	112	21

reports the chemical composition and the variability of the main components described in the *H. perforatum* flowers.

Analysis and standardisation of the extracts

The analysis of the components of St. John's Wort is complex because of the important diversity of constituents encountered. For a long period, the extracts were generally standardised in hypericin or hypericin-like substances. This red pigment was indeed considered as the undiscussed active ingredient and was the target of all analytical methods reported in literature [3].

Hypericin, thanks to its specific chromophore, was detected in the extracts by various colorimetric reactions, thin layer chromatography (TLC), gas liquid chromatography (GLC) or high performance liquid chromatography (HPLC) techniques [33, 73–75]. In the German pharmacopoeia, a determination of hypericin and derivatives is prescribed using a spectrophotometric assay at 500 nm.

More recently, different HPLC methods have been developed for the determination of all the constituents endowed with biological activities. Indeed, since the role of hypericin in the antidepressant activity of the extract has been much debated, the biological effects of *H. perforatum* are now mainly considered to arise, rather than from the presence of a single constituent, from the whole mixture of the metabolites.

In this respect, Brolis et al. [76] have described an HPLC-UV method using a wide pore RP-18 column. Chlorogenic acid (Fig. 6d), quercetin (Fig. 4a), quercitrin (Fig. 4b), isoquercitrin (Fig. 4c), rutin (Fig. 4d), hyperoside (Fig. 4e), I3,II8-biapigenin (Fig. 4f), pseudohypericin (Fig. 1b), hypericin (Fig. 1a), hyperforin (Fig. 2a) and adhyperforin (Fig. 2b) were separated by an aqueous phosphoric acid–acetonitrile–methanol gradient within 50 minutes.

The quantification of the above constituents was performed using rutin as an external standard. Another alternative HPLC method aimed at the quantification of the naphthodianthrones and the two phloroglucinols was reported by Poutaraud et al. [77], which carefully optimized the extraction procedure (water:ethanol; 40:60; 1 h; 80 °C in the dark) for the best compromise between extraction time and high recovery.

The major components of eight different batches of commercially available dry extracts of *H. perforatum* were also recently determined by a combination of analytical methods [72]. They include the quantification of flavonoids and phloroglucinols but also a wide variety of by-products that may act partially as co-effectors and affect the technological properties of the extracts. Thus 60–70% of the compounds of the *H. perforatum* dry extracts were efficiently quantified. The contents of hyperforin, hypericin and flavonoids were found to be in the range of 1.3–3.9%, 0.19–0.30%, and 4.8–11.4%, respectively. Water-soluble sugars were analysed by HPLC with refractive index detection. Native fructose, glucose and sucrose, as well as lactose added during the processing of the extracts, were determined. The total sugar content in the dry herbal extracts varied from 19–25% by weight. In addition, citric acid (0.9–2.3%) and malic acid (2.3–3.1%) were determined by HPLC, tannins (6.2–9.0%) and total ash (4.9–8.4%) were quantified according to the methods described in the *European Pharmacopoeia*, and the content of the total protein (3.9–8.3%) was estimated by elemental analysis [72].

A rapid method for the simultaneous determination of the six major naphthodianthrones and phloroglucinols by LC/ESI-MS/MS was also recently reported [16]. This method, based on multiple dissociation reaction monitoring (MRM), allowed the analysis of hypericin, protohypericin, pseudohypericin and protopseudophypericin in less than 5 min with lower level of quantification of 0.5 ng/ml for hyperforin and 2 ng/ml for hypericin.

References

1 Verotta L (2003) *Hypericum perforatum*, a source of neuroactive lead structures. *Current Topics Med Chem* 3: 187–201
2 Barnes J, Anderson LA, Phillipson JD (2001) St John's Wort (*Hypericum perforatum* L.): a review of its chemistry, pharmacology and clinical properties. *J Pharm Pharmacol* 53: 583–600
3 Bombardelli E, Morazzoni P (1995) *Hypericum perforatum*. *Fitoterapia* 56: 43–68
4 Roth L (1990) *Hypericum, Hypericin: Botanik, Inhaltsstoffe, Wirkung*. Vol Ecomed, Landsberg/Lech
5 Buchner A (1830) Buchner's Report. *Pharm* 34: 217
6 Siersch E (1927) Anatomie und Mikrochemie der Hypericum Drüsen (Chemie des Hypericins). *Plants* 3: 481
7 Brockmann H, Sanne W (1953) Pseudohypericin, ein neuer Hypericum-Farbstoff. *Naturwissenschaften* 40: 461
8 Brockmann H (1957) Photodynamisch wirksame Pflanzenfarbstoffe. *Fortschritt Chem Org Naturstoff* 14: 141–177
9 Brockmann H, Franssen U, Spitzner D, Augustiniak H (1974) Zur Isolierung und Konstitution des Pseudohypericins. *Tetrahedron Lett* 15: 1991–1994

10 Brockmann H, Haschad MN, Maier K, Pohl F (1939) Über das Hypericin, den photodynamisch wirksamen Farbstoff aus *H. perforatum*. *Naturwissenschaften* 27: 550
11 Brockmann H, Kluge F (1951) Zur Synthese des Hypericins. *Naturwissenschaften* 38: 141
12 Brockmann H, Sanne W (1957) Zur Kenntnis des Hypericins und Pseudohypericins. *Chem. Berichte* 90: 2480–2491
13 Brockmann H, Spitzner D (1975) Die Konstitution des Pseudohypericins. *Tetrahedron Lett* 16: 37–40
14 Cameron D, Raverty WD (1976) Pseudohypericin and other phenanthroperlyne quinones. *Aust J Chem* 29: 1523–1533
15 Mathis C, Ourisson G (1963) Etude chimio-taxonomique du genre *hypericum* I. Repartition de l'hypericine. *Phytochemistry* 2: 157–171
16 Tolonen A, Uusitalo J, Hohtola A, Jalonen J (2002) Determination of naphthodianthones and phloroglucinols from *Hypericum perforatum* extracts by liquid chromatography/tandem mass spectrometry. *Rap Comm Mass Spectrom* 16: 396–402
17 Freeman D, Kapinus E, Lavie D, Lavie G, Meruelo D, Mazur Y (1994) Acidic properties of hypericin and its octahydroxy analog in the ground and excited-states. *J Chem Soc Chem Commun* 7: 891–892
18 Lavie G, Mazur Y, Lavie D, Meruelo D (1995) The chemical and biological properties of Hypericin: a compound with a broad spectrum of biological activities. *Med Res Rev* 15: 111–119
19 Suzuki O, Katsumata Y, Oya M, Bladt S, Wagner H (1984) Inhibition of monoamine oxydase by hypericin. *Planta Med* 50: 272–274
20 Bladt S, Wagner H (1993) MAO-Hemmung durch Fraktionen und Inhaltsstoffen von *Hypericum*-Extrakt. *Nervenheilkunde* 12: 349–352
21 Yang KC, Prusti RK, Walker EB, Song P-S, Watanabe M, Furuya M (1986) Photodynamic action in *Stentor coeruleus* sensitized by endogenous pigment stentorin. *Photochem Photobiol* 43: 305–310
22 Barnard DL, Huffman JH, Wood SG (1990) Characterization of the anti human cytomegalovirus (HCMV) activity of three anthraquinone compounds 30th Interscience Conference on Antimicrobial Agents and Chemotherapy, Atlanta, October 21–24: 1093
23 Hudson JB, Lopez-Bazzocchi I, Towers GHN (1991) Antiviral activities of *Hypericum Antivir Res* 15: 101–112
24 Moraleda G, Wu TT, Jilbert AR, Aldrich CE, Condreay LD, Larsen SH, Tang JC, Colacino JM, Mason WS (1993) Inhibition of duck hepatitis B virus replication by hypericin. *Antivir Res* 20: 235–247
25 Tang J, Colacino JM, Larsen SH, Spitze W (1990) Virucidal activity of *Hypericum* against enveloped and non-enveloped DNA and RNA virus. *Antivir Res* 13: 313–325
26 Kraus GA, Pratt D, Tossberg J, Carpenter S (1990) Antiretroviral activity of synthetic hypericin and related analogs. *Biochem Biophys Res Commun* 172: 149–153
27 Lavie G, Valentine F, Levin B, Mazur Y, Gallo G, Lavie D, Weiner D, Meruelo D (1989) Studies of the mechanism of action of the antiretroviral agents hypericin and pseudohypericin. *Proc Natl Acad Sci USA* 86: 5963–5967
28 Lenard J, Rabson A, Vanderoef R (1993) Photodynamic inactivation of infectivity of human immunodeficiency virus and enveloped viruses using hypericin and rose bengal: inhibition of fusion and syncytia formation. *Proc Natl Acad Sci USA* 90: 158–162
29 Lopez-Bazzocchi I, Hudson JB, Towers GHN (1991) Antiviral activity of the photoactive plant pigment hypericin. *Photochem Photobiol* 54: 95
30 Meruelo D, Lavie G, Lavie D (1988) Therapeutic agents with dramatic antiretroviral activity and little toxicity at effective doses: aromatic polycyclic diones Hypericin and Pseudohypericin. *Proc Natl Acad Sci USA* 85: 5230–5234
31 Vlietinck AJ, De Bruyne T, Apers S, Pieters LA (1998) Plant-derived lead compounds for chemotherapy of human immunodeficiency virus (HIV) infection. *Planta Med* 64: 97–109
32 Gulick RM, McAuliffe V, Holden-Wiltse J, Crumpacker C, Liebes L, Stein DS, Meehan P, Hussey S, Forcht J, Valentine FT (1999) Phase I studies of hypericin, the active compound in St. John's Wort, as an antiretroviral agent in HIV-infected adults – AIDS clinical trials group protocols 150 and 258. *Ann Intern Med* 130: 510–514
33 Vanhaelen M, Vanhaelen-Fastre R (1983) Quantitative determination of biologically active constituents in medicinal plant crude extracts by thin layer chromatography-densitometry. *J Chromatogr* 281: 263–271

34 Gurevich AI, Dobrynin VN, Kolosov MN, Popravko SA, Ryabova ID, Chernov BK, Derbents NA, Aizenman BE, Garaguly AD (1971) Hyperforin, an antibiotic from *Hypericum perforatum* L. *Antibiotiki* 16: 510–513
35 Roz N, Mazur Y, Hirshfeld A, Rehavi M (2002) Inhibition of vesicular uptake of monoamines by hyperforin. *Life Sciences* 71: 2227–2237
36 Bystrov NS, Chernov BK, Dobrynin VN, Kolosov MN (1975) Structure of hyperforin. *Tetrahedron Lett* 32: 2791–2794
37 Bystrov NS, Dobrynin VN, Kolosov MN, Chernov BK, Chervin, II, Yakovlev GI (1975) Structure of chromophoric moiety of hyperforin. *Doklady Akademii Nauk Sssr* 225: 1327–1328
38 Bystrov NS, Dobrynin VN, Kolosov MN, Chernov BK, Chervin, II, Yakovlev GI (1976) Stoichiometry of antibiotic hyperforin. *Doklady Akademii Nauk Sssr* 226: 338–341
39 Bystrov NS, Gupta SR, Dobrynin VN, Kolosov MN, Chernov BK, Chervin, II, Yakovlev GI, Aizenman BE, Derbentseva NA (1976) Structure of antibiotic hyperforin. *Doklady Akademii Nauk Sssr* 226: 88–90
40 Brondz I, Greibrokk T, Groth P, Aasen AJ (1983) The absolute-configuration of hyperforin, an antibiotic from *Hypericum-perforatum* L, based on the crystal-structure determination of its *para*-bromobenzoate ester. *Acta Chem Scan* 37: 263–265
41 Maisenbacher P, Kovar KA (1992) Adhyperforin – a homolog of hyperforin from *Hypericum-perforatum. Planta Med* 58: 291–293
42 Shan MD, Hu LH, Chen ZL (2001) Three hyperforin analogues from *Hypericum perforatum. J Nat Prod* 64: 127–130
43 Fuzzati N, Gabetta N, Strepponi I, Villa F (2001) High-performance liquid chromatography–electrospray ionization mass spectrometry and multiple mass spectrometry studies of hyperforin degradation products. *J Chromatogr A* 926: 187–198
44 Trifunovic S, Vajs V, Macura S, Juranic N, Djarmati Z, Jankov R, Milosavljevic S (1998) Oxidation products of hyperforin from *Hypericum perforatum. Phytochemistry* 49: 1305–1310
45 Vajs V, Vugdelija S, Trifunovic S, Karadzic I, Juranic N, Macura S, Milosavljevic S (2003) Further degradation products of hyperforin from *Hypericum perforatum* (St. John's Wort). *Fitoterapia* 74: 439–444
46 Verotta L, Appendino G, Belloro E, Jakupovic J, Bombardelli E (1999) Furohyperforin, a prenylated phloroglucinol from St. John's Wort (*Hypericum perforatum*). *J Nat Prod* 62: 770–772
47 Verotta L, Appendino G, Belloro E, Jakupovic J, Bombardelli E (2000) Hyperforin analogues from St. John's Wort (*Hypericum perforatum*). *J Nat Prod* 63: 412–415
48 Wolfender J-L, Verotta L, Fuzzati N, Strepponi I, Hostettmann K (2003) Structural investigations of isomeric oxidised forms of hyperforin by LC/NMR and LC/MSn. *Phytochem Anal* 14: 290–297
49 Verotta L, Appendino G, Belloro E, Bianchi F, Sterner O, Lovati M, Bombardelli E (2002) Synthesis and biological evaluation of hyperforin analogues. Part I. Modification of the enolized cyclohexanedione moiety. *J Nat Prod* 65: 433–438
50 Jerzmanowska Z (1937) Über das Hypericin, ein Glukosid von *Hypericum perforatum* L. *Wiadomusci Farm* 64: 527
51 Sprecher P, Casparis F, Müller HJ (1946) Untersuchungen über natürliche Flavonolglykoside. *Pharm Acta Helvetiae* 21: 341
52 Dorossiev I (1985) Determination of flavonoids in *Hypericum perforatum. Pharmazie* 40: 585–586
53 Berghofer R, Holzl J (1987) Biflavonoids in *Hypericum-perforatum*.1. Isolation of I3,Ii8-Biapigenin. *Planta Med* 53: 216–217
54 Berghofer R, Holzl J (1989) Biflavonoids from *Hypericum-perforatum* L.2. Isolation of I3',Ii8-Biapigenin (Amentoflavone) from *Hypericum perforatum. Planta Med* 55: 91
55 Demisch L, Hölzl J, Golnik B, Kaczmarczyk P (1989) Identification of selective MAO-type-A inhibitors in *Hypericum perforatum* L. (Hyperforat®). *Pharmacopsychiatry* 22: 194
56 Melzer R, Fricke U, Holzl J (1991) Vasoactive properties of procyanidins from *Hypericum- perforatum*-L in isolated porcine coronary-arteries. *Arzneimittel-Forschung/Drug Research* 41: 481–483
57 Akhtardzhiev K, Kitanov G (1974) Catechin composition of *Hypericum perforatum. Farmatsiya* 24: 17–20
58 Gunatilaka AAL, Balasubramaniam S, Kumar V (1979) Studies on medicinal and related plants of Sri-Lanka.1. 2,3- dimethoxyxanthone from *Hypericum-mysorense. Phytochemistry* 18: 182–183

59 Suzuki O, Katsumata Y, Oya M, Chari VM, Klapfenberger R, Wagner H, Hostettmann K (1981) Inhibition of type A and type B monoamine oxidases by naturally occurring xanthones. *Planta Med* 42: 17–21
60 Sparenberg B, Demisch L, Hölzl J (1993) Untersuchungen über antidepressive Wirkstoffe von Johanniskraut. *J Pharm Ztg Wiss* 6: 50–54
61 Ollivier B, Balansard G, Maillard C, Vidal E (1985) Separation and identification of phenolic-acids by HPLC and UV spectral-analysis – Application to parietary (*Parietaria officinalis* L) and to Saint-Johns Wort (*Hypericum perforatum* L). *J Pharm Bel* 40: 173–177
62 Mathis C, Ourisson G (1964) Etude chimio-taxonomique du genre *Hypericum* II. Identification de constituants de diverses huiles essentielles d'*Hypericum*. *Phytochemistry* 3: 115–131
63 Mathis C, Ourisson G (1964) Etude chimio-taxonomique du genre *Hypericum* III. Répartition des carbures saturés et des monoterpènes dans les huiles essentielles d'*Hypericum*. *Phytochemistry* 3: 133–141
64 Mathis C, Ourisson G (1964) etude chimio-taxonomique du genre *Hypericum* IV. Répartition des sesquiterpenes, des alcools monoterpeniques et des aldehydes saturés dans les huiles essentielles d'*Hypericum*. *Phytochemistry* 3: 377–378
65 Schwob I, Bessiere JM, Viano J (2002) Composition of the essential oils of *Hypericum perforatum* L. from southeastern France. *CR Biologies* 325: 781–785
66 Mimica-Dukic N, Ivancev-Tumbas I, Igic R, Popovic M, Gasic O (1997) The content and composition of the essential oil of *Hypericum perforatum* from Serbia. *Pharm Pharmacol Let* 8: 26–28
67 Hölzl J, Ostrowski E (1987) Johanniskraut (*Hypericum perforatum* L.). HPLC-Analyse der wichtigen Inhaltstoffe und deren Variabilität in einer Population. *Dtsch Apoth Ztg* 127: 1227–1230
68 Brondz I, Greibrokk T, Aasen AJ (1983) Normal-alkanes of *Hypericum perforatum* – a revision. *Phytochemistry* 22: 295–296
69 Brondz I, Greibrokk T (1983) N-l-alkanols of *Hypericum perforatum*. *J Nat Prod* 46: 940–941
70 Commision_E_of_the_German_Federal_Public_Health_Departement (1984) *Monograph Hyperici herba, Johanniskraut*, Berlin
71 Berghofer R (1987) Dissertation, Universität Marburg
72 von Eggelkraut-Gottanka SG, Abu Abed S, Muller W, Schmidt PC (2002) Quantitative analysis of the active components and the by-products of eight dry extracts of *Hypericum perforatum* L. (St John's Wort). *Phytochem Anal* 13: 170–176
73 Holz J, Ostrowski E (1987) *Dtsch Apoth Ztg* 23
74 Kartnig T, Göbel I (1992) Determination of hypericin and pseudohypericin by thin-layer chromatography-densitometry. *J Chromat* 609: 423–426
75 Southwell IA, Campbell MH (1991) Hypericin content variation in *Hypericum perforatum* in Australia. *Phytochemistry* 30: 475–478
76 Brolis M, Gabetta B, Fuzzati N, Pace R, Panzeri F, Peterlongo F (1998) Identification by high-performance liquid chromatography-diode array detection-mass spectrometry and quantification by high performance liquid chromatography-UV absorbance detection of active constituents of *Hypericum perforatum*. *J Chromatogr A* 825: 9–16
77 Poutaraud A, Lobstein A, Girardin P, Weniger B (2001) Improved procedure for the quality control of *Hypericum perforatum* L. *Phytochem Anal* 12: 355–362

Edited by W. E. Müller

St. John's Wort and its active principles in depression and anxiety – A critical analysis of receptor binding studies

Marco Gobbi and Tiziana Mennini

Istituto di Ricerche Farmacologiche "Mario Negri", Via Eritrea 62, 20157 Milano, Italy

Introduction

The herb *Hypericum perforatum* (popularly called St. John's Wort, SJW) has been known for a long time for its putative medicinal properties including wound-healing, diuretic, antibiotic and antiviral effects [1]. Nowadays, the therapeutic use of alcoholic extracts of SJW tends to be concentrated on some central nervous system (CNS) disorders. The main indication, supported by a number of randomised clinical trials (reviewed in [2]), is for the treatment of less severe forms of depressive disorders, as an alternative to the classic antidepressants with a favourable side effect profile. SJW extracts show antidepressant-like properties in behavioural models in rodents [3–8] and have also been proposed – and in some countries they are licensed – for the treatment of anxiety and sleep disorders, although such indications are only supported by preclinical evidence [9–12]. Preclinical studies also suggest potential uses of SJW extracts in cases of alcohol abuse [13, 14].

Hydroalcoholic SJW extracts contain several natural products [15] belonging to the following main chemical classes: flavonoids (including flavonol glycosides such as rutin, quercetin and quercitrin; and biflavones such as biapigenin and amentoflavon); proanthocyanidins; naphthodianthrones (including hypericin and pseudohypericin) and acylphloroglucinols (including hyperforin and adhyperforin). Studies in animal models reported antidepressant-like properties of flavonoids [16, 17], of hypericin and pseudohypericin [18], and of hyperforin [19–21].

The mechanism(s) of action of the central effects of SJW extracts are still under investigation. Different hypothesis have been formulated on the basis of the results of *in vitro*, *ex vivo* and *in vivo* studies, using different extracts and some pure constituents. However, only *in vitro* receptor binding studies will be reviewed here and their interpretations and implications discussed.

Receptor binding assays

In vitro receptor binding studies, particularly the so-called "competition" binding assays, are widely used to determine whether an unlabelled compound interacts directly with a receptor of interest, and to measure the affinity of this interaction.

Briefly, the basic technique involves the incubation of a receptor preparation (from brain membranes or from cells expressing recombinant receptors) with a ligand (usually radiolabelled) that has high selectivity for the receptor. Once equilibrium is reached the radioligand bound to the receptor is separated from the free ligand (by filtration or centrifugation) and quantified. In a typical competition assay, the radioligand binding is measured in the presence of increasing concentrations of an unlabelled compound, to determine the IC_{50}, the concentration inhibiting the specific binding by 50%. This indicates the potency of this compound's effect on the radioligand binding and, if the inhibition is competitive, the compound's affinity for the receptor under investigation. Usually the "affinity" is indicated by the Ki value, which is the equilibrium dissociation constant: the lower the Ki, the higher the affinity. For a competitive inhibitor, Ki is calculated from the IC_{50} using the equation of Cheng and Prusoff [22], which takes into account the dissociation constant of the radioligand (Kd) and its concentration [L]: $Ki = IC_{50}/(1 + [L]/Kd)$.

The measure of the (relative) affinities of a given compound for different receptors is widely used as an easier and faster method to characterize the compound's pharmacology, suggesting therapeutic applications in the case of new compounds or helping clarify the mechanism of action.

In vitro binding data

Competition binding assays have tested the effects of different SJW extracts and many of the constituents, on binding to almost all central neurotransmitter receptors. Before reviewing these data, however, some points need to be considered for interpretation purposes.

In order to understand whether the affinity of a compound for a given receptor, determined *in vitro*, underlies the central effects of interest (i.e., the antidepressant effect), one needs to know whether high enough concentrations of that compound can actually reach the brain *in vivo*. For example, the affinity of hypericin for adrenergic β1 receptors (i.e., the concentrations occupying half the receptors) is 4.4 μM [23], indicating that hypericin can bind these receptors *in vivo* only if micromolar concentrations are achieved in the brain. Only in this case would the interaction with adrenergic β1 receptors be relevant for the pharmacological effects of hypericin or hypericin-containing SJW extracts.

It follows that the affinity values determined *in vitro* are meaningful only in the presence of pharmacokinetic data showing the brain levels of the single constituents. For example, amentoflavone interacts with nanomolar affinities with benzodiazepine and opioid δ receptors [23–25], but we cannot even guess at the relevance of these interactions (do they play a role in the antidepressant effect of SJW extract?) unless brain amentoflavone levels are measured after effective doses of the extract. Although it was recently reported that amentoflavone is able to pass the blood–brain barrier *in vitro* by passive diffusion [26], a previous study had shown that amentoflavone does not bind brain benzodiazepine receptors after i.v. administration in the mouse *in vivo* [24].

For these reasons, we mainly focus here on the affinity values for hypericin and hyperforin, since these are the only components for which estimates of brain levels are available (see below). Total hypericins (i.e., the sum of hypericin, pseudohypericin and their protocompounds) usually amount to about 0.3% of the SJW extract. The content of hyperforin is highly variable, ranging from 1–5%, because of its instability to light and air [27]. Hyperforin plus hypericin make up less than 10% of the total SJW extract. These considerations make it even more difficult to interpret the "affinity" values from *in vitro* binding studies with the total SJW extracts [6, 28, 29], because we have no idea which component is responsible for the effects on the ligand binding and whether that component can enter the brain.

Brain levels of hypericin and hyperforin

In humans, after single or multiple pharmacological doses of a SJW extract (LI-160), the maximal *plasma* concentrations of hypericin and pseudohypericin were always below 40 µg/L, i.e., lower than 80 nM [30]. As regards the *brain* hypericin concentrations, the only indication came from a study using radiolabeled hypericin in mice, showing that they amount to 8–40% of the plasma hypericin concentration [31]. The cerebrospinal fluid penetration of hypericin in nonhuman primates was less than 1% of plasma levels [32]. Maximal *plasma* concentrations of hyperforin were below 400 µg/L (about 800 nM) after single or multiple pharmacological doses of a SJW extract (WS-5572), in human and rat [21, 33]. However, preliminary brain-to-plasma distribution studies suggested very poor passage of the blood–brain barrier with brain hyperforin concentrations being only 4% of the plasma concentrations [21]. The hyperforin concentration in the brain of mice treated with pharmacological doses of WS-5572 was 15.8 ng/g (about 30 nM, assuming 1 g brain tissue equivalent to 1 mL water) [34].

In conclusion, pharmacokinetic data indicate that after pharmacologically effective doses of SJW extracts, the brain concentration of hypericin and hyperforin is below 50 nM. It can therefore be assumed that these two constituents will *not* interact with those central neurotransmitter receptors for which their Ki exceed, conservatively, 500 nM (with this Ki value 50 nM of the compound will occupy, at equilibrium, less than 10% of receptors).

Most binding studies report IC_{50} instead of Ki values. The IC_{50} value is proportional to Ki but also depends on experimental conditions such as the affinity of the radioligand used and its concentration. Since binding studies usually employ a radioligand concentration equal to or lower than its Kd, the Cheng and Prusoff equation calculates that a Ki of 500 nM (i.e., the threshold identified above) corresponds to an IC_{50} of 1,000 nM or lower.

Affinity of hypericin and hyperforin for central neurotransmitter receptors

Ki/IC50 values *higher* than the thresholds above (0.5 and 1 μM respectively), were reported for hyperforin and/or hypericin on the following central neurotransmitter receptors:

- Monoamine transporters [6, 23, 35, 36].
- Serotonin receptors (all subtypes, with the exception of the 5-HT4 subtype for which no data are available) [6, 23, 29, 35, 36].
- Dopamine receptors, all subtypes [23, 36] with the exception of hypericin on the DA3 subtype, see below.
- Adrenergic receptors (hypericin only on all subtypes) [23, 36].
- Acetylcholine receptors. In one study hypericin showed an IC_{50} of about 1 μM on rat brain cortex muscarinic receptors (subtypes not measured) [36], but this finding was not confirmed when evaluating its effect on recombinant muscarinic 1–5 subtypes ($IC_{50} > 10$ μM) [23]. No data are available for hyperforin.
- Histamine receptors (H1 subtype) [23, 36].
- GABA-A receptors [23, 35, 36].
- Benzodiazepine receptors [23, 25, 35, 36].
- Glutamate-NMDA receptors [23, 36].
- Glutamate-PCP receptors [23, 36].
- Opioid μ receptors. Hyperforin and hypericin were quite active in one study (IC_{50} 0.4 and 1 μM, respectively) [29] but this finding was not confirmed later ($IC_{50} > 10$ μM) [23].
- Opioid κ receptors. In one study hyperforin and hypericin had IC_{50} of 1 and 3 μM, respectively [29] but this finding was not confirmed later ($IC_{50} > 10$ μM) [23].
- Opioid δ receptors. In one study hyperforin and hypericin had IC_{50} of 0.5 and 4 μM, respectively [29] but this finding was not confirmed later ($IC_{50} > 10$ μM) [23].
- Sigma (σ) receptors. No effect of hyperforin [35] whereas hypericin had an IC_{50} of about 1–3 μM [35, 36]. However, the inhibitory effect of hypericin was light-dependent, being much lower when the binding assay was carried out in the dark [35].
- Neuropeptide-Y receptors (Y1-2 subtypes) [35, 36].

- Neurokinin-1 receptors [36, 37].
- Vasopressin receptors (V1-3 subtypes) [23, 36].
- Bradykinin-2 receptors, hypericin only [36].
- Cholecystokinin-A receptors, hypericin only [36].
- Endothelin-A receptors, hypericin only [36].
- Angiotensin-1 receptors, hypericin only [36].
- Glucocorticoids receptors, hypericin only [36].

The only reported interactions of hypericin/hyperforin with sufficiently high affinity were:

- Hypericin with rDA3 receptors, with a Ki of 34 nM [23].
- Hypericin with CRF-1 receptors, with an IC_{50} of 300 nM [29].

The dopaminergic system is involved in the pharmacological effects of SJW extracts as indicated by the fact that sulpiride (a DA_2/DA_3 receptor antagonist) and haloperidol (D_2/D_4 receptor antagonist) completely antagonised the effects of the extracts in the forced swimming test in rats [3, 38]. Sulpiride also antagonised the anti-immobility effect of solubilised hypericin [18]. These effects of the DA antagonists might be due to inhibition of hypericin binding to DA3 receptors (note however that behavioural data mainly suggest the involvement of DA2 receptors) but they could also be due to antagonism of DA itself, whose extracellular concentrations are enhanced by treatment with SJW extracts [39, 40]. Various antidepressant drugs, including those with no direct effect on central DA mechanisms, enhance the sensitivity of postsynaptic DA receptors, including DA3 receptors [41, 42] in the mesolimbic system [43]. There is no evidence yet that ligands to DA3 receptors have antidepressant activity.

CRF1 receptors are major determinants in the regulation of the hypothalamic–pituitary–adrenal axis, whose dysregulation is thought to play a causal role in the development and course of depression [44]. Clinical and preclinical data suggest that unrestrained secretion of CRF in the CNS produces several signs and symptoms of depression and anxiety disorders through continuous activation of CRF1 receptors. As a consequence, selective CRF1 receptor antagonists are being developed as potential anxiolytics/antidepressants [45].

The interaction of SJW constituents with CRF1 receptors might therefore be important. As regards hyperforin, preliminary binding studies showed it was inactive on recombinant human receptors at concentrations up to 10 μM [29]. Functional *in vitro* studies, however, showed that hyperforin inhibits the CRF-induced cAMP accumulation in CHO-K1 cells expressing CRF-1 receptors with an IC_{50} of about 1 μM [46]. Since this antagonism is noncompetitive it is probably due to hyperforin interacting with ion channels or with some other membrane-related effect, as has been described at similar concentrations of this compound [47]. Hypericin inhibited CRF-induced cAMP accumulation with a Kb of 930 nM in a competitive manner [46], in partial agreement with its relatively high affinity for CRF1 receptors (300 nM) [29]. However, these hypericin concentrations are quite high, very near the "conservative" thresholds identified above.

Concluding remarks

The main conclusion that can be drawn from *in vitro* binding data is that the central effects of SJW extracts are not due to hypericin or hyperforin interacting *directly* with most of the central neurotransmitter transporters and receptors, because the concentrations required for these interactions far exceed those found in the brain after pharmacologically effective doses of the extract. The only exceptions might be the relatively high affinities reported for hypericin on DA3 or CRF-1 receptors.

Since behavioural studies in animal models consistently suggest that hyperforin and hypericin are involved in the antidepressant-like effects of SJW extracts [18–21] it can be speculated that hyperforin/hypericin:

1. Act peripherally level, e.g., on the cytokine metabolism [37]
2. Interact with high affinity with a central target not yet evaluated in binding assays (new mechanism of action?)
3. Form metabolites, not yet detected and measured in the brain, that interact with high affinity with the classic receptors involved in depression and anxiety
4. Induce bioactive endogenous molecules through an unknown mechanism
5. Act synergistically with other constituents, so that various relatively weak effects result in the overall pharmacological effect [29].

Regarding points 3 and 4: it must be considered that an indirect effect on sigma receptor could be involved in the mechanism of SJW actions. In spite of the lack of affinities of hypericin/hyperforin for sigma receptors *in vitro*, pretreatment of rats with pharmacologically active doses of SJW extracts reduced ligand binding to sigma receptors, measured *ex vivo* [48, 49]. Consistent with this possibility is the finding that rimcazole (a $sigma_1$ receptor antagonist) counteracts the antidepressant effects of the SJW extract, evaluated with the forced swimming test in rats [8], and that agonists at $sigma_1$ receptors are active in antidepressant models in rats [50].

Regarding point 5: flavonoids are also likely to play a role in the central effects of SJW extracts [16, 17], and amentoflavone interacts with high affinity with benzodiazepine receptors [23–25]. Unfortunately, this last finding cannot be adequately interpreted because of the lack of pharmacokinetic data on the brain levels of this important constituent after treatment with SJW extracts.

In summary, different experimental approaches (*in vitro*, *ex vivo* and *in vivo*) are required to clarify the difficult question of the active principle(s) and the mechanism(s) of action underlying the pharmacological effects of a complex mixture such as SJW extracts [51]. It is very important to interpret the results with each of these approaches correctly. Data from *in vitro* binding assays have often been over interpreted, mainly because they have been considered without taking into account the (scant) pharmacokinetic data available. Additional pharmacokinetic data on the brain concentrations of other constituents and/or metabolites are therefore required for a more meaningful analysis of *in vitro* receptor binding data which, it is to be hoped, should enable us to identify the active principle and its target in the brain.

References

1 Bombardelli E, Morazzoni P (1995) *Hypericum perforatum. Fitoterapia* 66: 43–68
2 Linde K, Mulrow CD (2003) St John's Wort for depression. *Cochrane Database Systematic Reviews* CD000448
3 Butterweck V, Wall A, Lieflander-Wulf U, Winterhoff H, Nahrstedt A (1997) Effects of the total extract and fractions of *Hypericum perforatum* in animal assays for antidepressant activity. *Pharmacopsychiatry* 30 (Suppl 2): 117–124
4 Ozturk Y (1997) Testing the antidepressant effects of *Hypericum* species on animal models. *Pharmacopsychiatry* 30 (Suppl 2): 125–128
5 Chatterjee SS, Noldner M, Koch E, Erdelmeier C (1998) Antidepressant activity of *Hypericum perforatum* and hyperforin: the neglected possibility. *Pharmacopsychiatry* 31 (Suppl 1): 7–15
6 Gobbi M, Dalla Valle F, Ciapparelli C, Diomede L, Morazzoni P, Verotta L, Caccia S, Cervo L, Mennini T (1999) *Hypericum perforatum* L. extract does not inhibit 5-HT transporter in rat brain cortex. *Naunyn Schmiedebergs Arch Pharmacol* 360: 262–269
7 Gambarana C, Ghiglieri O, Tolu P, De Montis MG, Giachetti D, Bombardelli E, Tagliamonte A (1999) Efficacy of an *Hypericum perforatum* (St. John's Wort) extract in preventing and reverting a condition of escape deficit in rats. *Neuropsychopharmacology* 21: 247–257
8 Panocka I, Perfumi M, Angeletti S, Ciccocioppo R, Massi M (2000) Effects of *Hypericum perforatum* extract on ethanol intake, and on behavioral despair: a search for the neurochemical systems involved. *Pharmacol Biochem Behav* 66: 105–111
9 Beijamini V, Andreatini R (2003) Effects of *Hypericum perforatum* and paroxetine in the mouse defense test battery. *Pharmacol Biochem Behav* 74: 1015–1024
10 Flausino OA Jr, Zangrossi H Jr, Salgado JV, Viana MB (2002) Effects of acute and chronic treatment with *Hypericum perforatum* L. (LI 160) on different anxiety-related responses in rats. *Pharmacol Biochem Behav* 71: 251–257
11 Vandenbogaerde A, Zanoli P, Puia G, Truzzi C, Kamuhabwa A, De Witte P, Merlevede W, Baraldi M (2000) Evidence that total extract of *Hypericum perforatum* affects exploratory behavior and exerts anxiolytic effects in rats. *Pharmacol Biochem Behav* 65: 627–633
12 Schulz H, Jobert M (1994) Effects of *Hypericum* extract on the sleep EEG in older volunteers. *J Geriatr Psychiatry Neurol* 7 (Suppl 1): S39–43
13 Perfumi M, Ciccocioppo R, Angeletti S, Cucculelli M, Massi M (1999) Effects of *Hypericum perforatum* extraction on alcohol intake in Marchigian Sardinian alcohol-preferring rats. *Alcohol Alcohol* 34: 690–698
14 Rezvani AH, Overstreet DH, Yang Y, Clark E Jr, (1999) Attenuation of alcohol intake by extract of *Hypericum perforatum* (St. John's Wort) in two different strains of alcohol-preferring rats. *Alcohol Alcohol* 34: 699–705
15 Nahrstedt A, Butterweck V (1997) Biologically active and other chemical constituents of the herb of *Hypericum perforatum* L. *Pharmacopsychiatry* 30 Suppl 2: 129–134
16 Butterweck V, Jurgenliemk G, Nahrstedt A, Winterhoff H (2000) Flavonoids from *Hypericum perforatum* show antidepressant activity in the forced swimming test. *Planta Med* 66: 3–6
17 Butterweck V, Christoffel V, Nahrstedt A, Petereit F, Spengler B, Winterhoff H (2003) Step by step removal of hyperforin and hypericin: activity profile of different *Hypericum* preparations in behavioral models. *Life Sci* 73: 627–639

18 Butterweck V, Petereit F, Winterhoff H, Nahrstedt A (1998) Solubilized hypericin and pseudohypericin from *Hypericum perforatum* exert antidepressant activity in the forced swimming test. *Planta Med* 64: 291–294
19 Chatterjee SS, Bhattacharya SK, Wonnemann M, Singer A, Muller WE (1998) Hyperforin as a possible antidepressant component of *Hypericum* extracts. *Life Sci* 63: 499–510
20 Gambarana C, Tolu PL, Masi F, Rinaldi M, Giachetti D, Morazzoni P, De Montis MG (2001) A study of the antidepressant activity of *Hypericum perforatum* on animal models. *Pharmacopsychiatry* 34 (Suppl 1): S42–44
21 Cervo L, Rozio M, Ekalle-Soppo CB, Guiso G, Morazzoni P, Caccia S (2002) Role of hyperforin in the antidepressant-like activity of *Hypericum perforatum* extracts. *Psychopharmacology (Berl)* 164: 423–428
22 Cheng Y, Prusoff WH (1973) Relationship between the inhibition constant (K1) and the concentration of inhibitor which causes 50 per cent inhibition (I50) of an enzymatic reaction. *Biochem Pharmacol* 22: 3099–3108
23 Butterweck V, Nahrstedt A, Evans J, Hufeisen S, Rauser L, Savage J, Popadak B, Ernsberger P, Roth BL (2002) *In vitro* receptor screening of pure constituents of St. John's Wort reveals novel interactions with a number of GPCRs. *Psychopharmacology (Berl)* 162: 193–202
24 Nielsen M, Frokjaer S, Braestrup C (1988) High affinity of the naturally-occurring biflavonoid, amentoflavon, to brain benzodiazepine receptors *in vitro. Biochem Pharmacol* 37: 3285–3287
25 Baureithel KH, Buter KB, Engesser A, Burkard W, Schaffner W (1997) Inhibition of benzodiazepine binding *in vitro* by amentoflavone, a constituent of various species of *Hypericum. Pharm Acta Helv* 72: 153–157
26 Gutmann H, Bruggisser R, Schaffner W, Bogman K, Botomino A, Drewe J (2002) Transport of amentoflavone across the blood-brain barrier *in vitro. Planta Med* 68: 804–807
27 Erdelmeier CA (1998) Hyperforin, possibly the major non-nitrogenous secondary metabolite of *Hypericum perforatum* L. *Pharmacopsychiatry* 31 (Suppl 1): 2–6
28 Cott JM (1997) *In vitro* receptor binding and enzyme inhibition by *Hypericum perforatum* extract. *Pharmacopsychiatry* 30 (Suppl 2): 108–112
29 Simmen U, Higelin J, Berger-Buter K, Schaffner W, Lundstrom K (2001) Neurochemical studies with St. John's Wort *in vitro. Pharmacopsychiatry* 34 (Suppl 1): S137–142
30 Brockmoller J, Reum T, Bauer S, Kerb R, Hubner WD, Roots I (1997) Hypericin and pseudohypericin: pharmacokinetics and effects on photosensitivity in humans. *Pharmacopsychiatry* 30 (Suppl 2): 94–101
31 Stock S, Holzl J (1991) Pharmacokinetic tests of [^{14}C]-labeled hypericin and psudohypericin from *Hypericum perforatum* and serum kinetics of hypericin in man. *Planta Med* 57 (Suppl 2): A61
32 Fox E, Murphy RF, McCully CL, Adamson PC (2001) Plasma pharmacokinetics and cerebrospinal fluid penetration of hypericin in nonhuman primates. *Cancer Chemother Pharmacol* 47: 41–44
33 Biber A, Fischer H, Romer A, Chatterjee SS (1998) Oral bioavailability of hyperforin from *Hypericum* extracts in rats and human volunteers. *Pharmacopsychiatry* 31 (Suppl 1): 36–43
34 Keller JH, Karas M, Muller WE, Volmer DA, Eckert GP, Tawab MA, Blume HH, Dingermann T, Schubert-Zsilavecz M (2003) Determination of hyperforin in mouse brain by High-Performance Liquid Chromatography/Tandem Mass Spectrometry. *Anal Chem* 75: 6084–6088
35 Gobbi M, Moia M, Pirona L, Morazzoni P, Mennini T (2001) *In vitro* binding studies with two *Hypericum perforatum* extracts – hyperforin, hypericin and biapigenin – on 5-HT6, 5-HT7, GABA(A)/benzodiazepine, sigma, NPY-Y1/Y2 receptors and dopamine transporters. *Pharmacopsychiatry* 34 (Suppl 1): S45–48
36 Raffa RB (1998) Screen of receptor and uptake-site activity of hypericin component of St. John's Wort reveals sigma receptor binding. *Life Sci* 62: PL265–270
37 Gobbi M, Moia M, Funicello M, Mennini T (2004) *In vitro* effects of hyperforin, a putative active principle of *Hypericum perforatum*, on IL6 release in different experimental models. *Planta Med* 70: 680–682
38 Calapai G, Crupi A, Firenzuoli F, Inferrera G, Squadrito F, Parisi A, De Sarro G, Caputi A (2001) Serotonin, norepinephrine and dopamine involvement in the antidepressant action of *Hypericum perforatum. Pharmacopsychiatry* 34: 45–49
39 Calapai G, Crupi A, Firenzuoli F, Costantino G, Inferrera G, Campo GM, Caputi AP (1999) Effects of *Hypericum perforatum* on levels of 5-hydroxytryptamine, noradrenaline and dopamine in the cortex, diencephalon and brainstem of the rat. *J Pharm Pharmacol* 51: 723–728

40 Di Matteo V, Di Giovanni G, Di Mascio M, Esposito E (2000) Effect of acute administration of *Hypericum perforatum*-CO_2 extract on dopamine and serotonin release in the rat central nervous system. *Pharmacopsychiatry* 33: 14–18
41 Lammers CH, Diaz J, Schwartz JC, Sokoloff P (2000) Dopamine D3 receptor gene expression in the shell of nucleus accumbens is increased by chronic antidepressant treatment. *Mol Psychiatry* 5: 229
42 Maj J, Dziedzicka-Wasylewska M, Rogoz R, Rogoz Z (1998) Effect of antidepressant drugs administered repeatedly on the dopamine D3 receptors in the rat brain. *Eur J Pharmacol* 351: 31–37
43 Garattini S, Samanin R (1988) Biochemical hypotheses on antidepressant drugs: a guide for clinicians or a toy for pharmacologists? *Psychol Med* 18: 287–304
44 Holsboer F (1999) The rationale for corticotropin-releasing hormone receptor (CRH-R) antagonists to treat depression and anxiety. *J Psychiatr Res* 33: 181–214
45 Zobel AW, Nickel T, Kunzel HE, Ackl N, Sonntag A, Ising M, Holsboer F (2000) Effects of the high-affinity corticotrophin-releasing hormone receptor 1 antagonist R121919 in major depression: the first 20 patients treated. *J Psychiatr Res* 34: 171–181
46 Simmen U, Bobirnac I, Ullmer C, Lubbert H, Berger Buter K, Schaffner W, Schoeffter P (2003) Antagonist effect of pseudohypericin at CRF1 receptors. *Eur J Pharmacol* 458: 251–256
47 Gobbi M, Mennini T (2001) Is St John's Wort a 'Prozac-like' herbal antidepressant? *Trends Pharmacol Sci* 22: 557–559
48 Pirona L, Funicello M, Ekalle Soppo CB, Cervo L, Bombardelli E, Morazzoni P, Mennini T, Gobbi M (2002) *Studi sul possibile meccanismo d'azione dell'iperico, una pianta medicinale con proprietà antidepressive. XIII Congresso della Società Italiana di Neuropsicofarmacologia,* Milan, Italy
49 Cervo L, Mennini T, Rozio M, Canetta A, Burbassi S, Guiso G, Pirona L, Riva A, Morazzoni P, Caccia S et al. (2004) Potential antidepressant properties of hyperforin-trimethoxybenzoate, a semisynthetic ester of hyperforin. *Eur Neuropsychopharmacol; in press*
50 Matsuno K, Kobayashi T, Tanaka MK, Mita S (1996) Sigma 1 receptor subtype is involved in the relief of behavioral despair in the mouse forced swimming test. *Eur J Pharmacol* 312: 267–271
51 Mennini T, Gobbi M (2003) The antidepressant mechanism of *Hypericum perforatum. Life Sci* 75: 1021–1027

St. John's Wort and its Active Principles in Depression and Anxiety
Edited by W. E. Müller

Effects on transmitter uptake and their cellular and molecular basis

Kristina Treiber and Walter E. Müller

Pharmakologisches Institut für Naturwissenschaftler, Biozentrum Universität Frankfurt, Marie-Curie-Str. 9, 60439 Frankfurt, Germany

The biological basis of depressive disorders is rather complex and not well understood, but a wealth of evidence points to a central pathophysiological as well as pharmacological role of serotonergic and noradrenergic neurotransmission. Accordingly, nearly all presently available antidepressant drugs influence the synaptic communication of the neurotransmitters serotonin and norepinephrine and to some extent of dopamine in the central nervous system (CNS), although different biochemical pathways are involved (Fig. 1). The majority of antidepressants lead, at least initially, to an increased availability of serotonin and norepinephrine at their respective synapses. This explains why it is quite convincible that the mode of action of St. John's Wort is also related to the synaptic communication of norepinephrine, serotonin, and dopamine. One possible biochemical mechanism of antidepressant activity is the inhibition of the intra- and extraneuronally located enzyme monoamine oxidase (MAO), a pathway which is exploited by the MAO inhibitors (Fig. 1). By this mechanism, the degradation of all three neurotransmitters is retarded and their concentration in the synaptic cleft increases. Earlier investigations, which assumed an inhibitory effect of St. John's Wort extract on MAO-A enzyme [1] and considered the extract as a herbal MAO inhibitor, could not be confirmed in later studies [2–7]. As an example, data from our laboratory are shown in Figure 2, where inhibition of MAO-A and MAO-B activity was only seen at very high concentrations, about two orders of magnitude above the concentrations needed to inhibit neuronal transmitter uptake (Fig. 3). Moreover, we observed no MAO inhibition *ex vivo* after acute or chronic treatment of mice and/or rats with St. John's Wort extract (data not shown). Thus, St. John's Wort is not a herbal MAO-inhibitor, as it is still indicated even in recent publications.

The most important mechanism of antidepressant activity goes back to the classic tricyclic antidepressants but is also valid for almost all "new" drugs. This mechanism is characterised by competitive inhibition of serotonin or/and norepinephrine transporter proteins, which return the neurotransmitter released into the synaptic cleft back into the presynaptic nerve terminals. Some old and new antidepressants are rather specific for only one transporter pro-

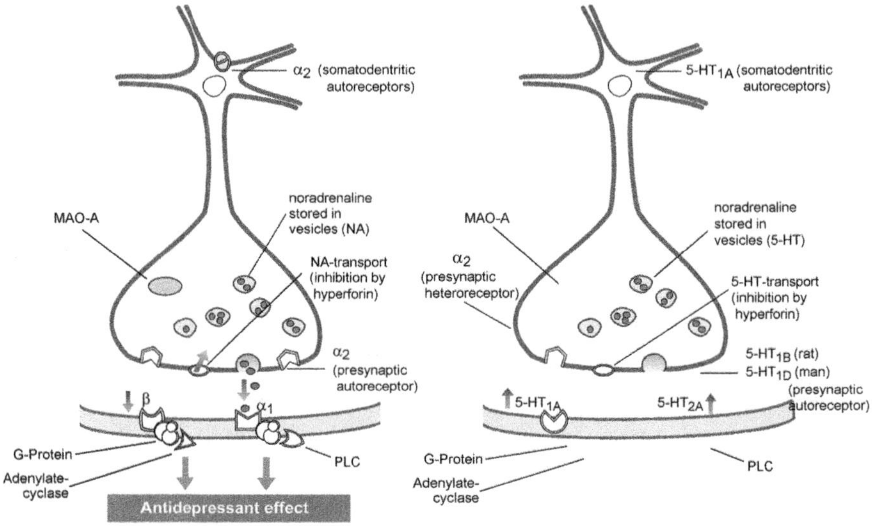

Figure 1. The biochemical mechanism of action of St. John's Wort extract exhibits similarities to that of other antidepressants. Many synthetic antidepressants lead to an increase in concentration of the two neurotransmitters noradrenaline and serotonin in the synapses (contact sites between nerve cells) of the brain by, at least initially, influencing different mechanisms (inhibition of neuronal noradrenaline or serotonin reuptake, inhibition of monoamine oxidase-A, inhibition of pre-synaptic α_2-receptors). As a consequence, adaptive changes in the post-synaptic receptor system take place, which occur over the same period as the antidepressant action. St. John's Wort also blocks both transport systems (mainly by means of the active ingredient hyperforin) and leads to changes in the β-, 5-HT1A- and 5-HT2A-receptors (see arrows).

tein, while other old and new drugs inhibit norepinephrine as well as serotonin transporters. However, none of the synthetic antidepressants inhibits the neuronal uptake of all three neurotransmitters serotonin, norepinephrine, and dopamine with comparable potencies like St. John's Wort extract does.

St. John's Wort extract is a broad spectrum inhibitor of neurotransmitter uptake

Already our initial findings indicate that St. John's Wort extract has a clear inhibitory effect on the synaptosomal uptake not only of serotonin and of norepinephrine, but also of dopamine (Fig. 3), GABA and L-Glutamate with rather similar potencies in mouse brain tissue (Tab. 1) [7]. These findings were confirmed in rat synaptosomes, primary cultures of rat neurons, and cultured rat cortical astrocytes [7–11]. The IC_{50} values of the extract as an inhibitor of neurotransmitter uptake are at least 100 times lower than those for the MAO-A or MAO-B inhibition (Tab. 1, Fig. 2). Importantly, upon removal of *Hypericum* extract, uptake was restored in neurons, thereby indicating that inhibition was not due to a toxic effect of *Hypericum* extract (Neary et al., 2001). Regarding

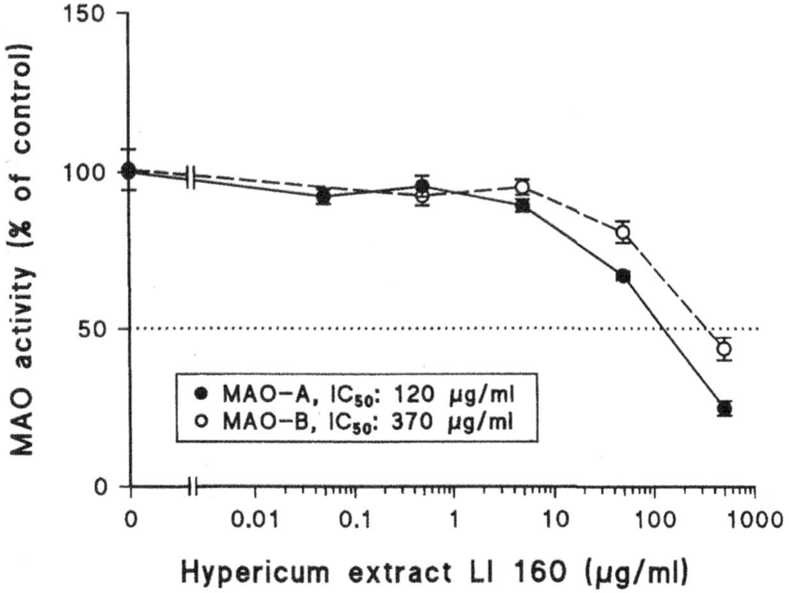

Figure 2. Inhibition of mouse brain MAO-A and MAO-B activities by *Hypericum* extract (2% hyperforin) *in vitro*. Data are from [7].

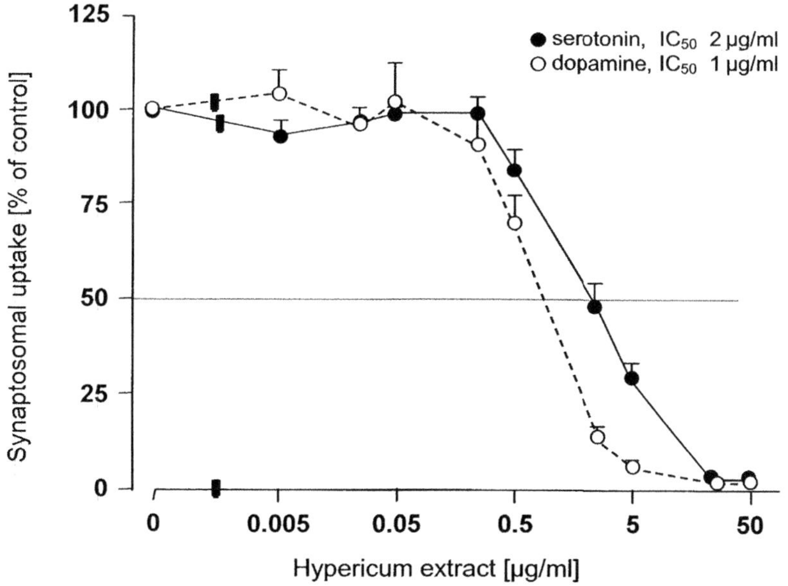

Figure 3. Inhibition of uptake of ^{3}H-serotonin and ^{3}H-dopamine into mouse cortex or rat striatum synaptosomes by *Hypericum* extract (2% hyperforin). Data are from [7].

Table 1. Mean inhibitory concentration (IC_{50}) of several antidepressants and of St. John's Wort extract (hyperforin content 2%) for the synaptosomal uptake of various neurotransmitters in mouse brain synaptosomes

Substance	Serotonin	Dopamine	Noradrenaline	GABA	L-Glutamate
IC_{50} (nmol/l)					
Imipramine	21	>1,000	21	>1,000	>1,000
Clomipramine	1	>1,000	14	>1,000	>1,000
Desipramine	207	>1,000	3	>1,000	>1,000
Citalopram	1	>1,000	>1,000	>1,000	>1,000
IC_{50} (μg/ml)					
St. John's Wort extract	2	1	5	1	11

Data are taken from [7] and [9].

the broad range of neurotransmitter uptake systems affected by St. John's Wort no other antidepressant has similar properties (Tab. 1).

In agreement with these findings on neuronal transmitter uptake *in vitro*, a number of authors found rather different changes in brain concentrations of norepinephrine, dopamine and serotonin or their respective metabolites after acute and chronic treatment of experimental animals with hypericum extract [12–15]. As an example, our data on the effects of the extract in the mouse brain after a single acute dose are given in Figure 4. For synthetic antidepressants, comparable data are also not uniform, since different effects are observed depending on the dose, the duration of treatment and the brain area investigated. In general however, these findings confirm that doses of St. John's Wort that are active in behavioural tests also cause changes in the above-mentioned neurotransmitter systems in the brain *in vivo*. Further evidence that St. John's Wort interferes with noradrenergic, serotonergic and dopaminergic neurotransmission *in vivo,* comes from alterations of β-adrenergic and serotonergic receptor densities after

Table 2. Inhibition of synaptosomal uptake systems by *Hypericum* extracts and hyperforin

	Hypericum extract (hyperforin content)			
	2%	39%	Hyperforin	
Uptake-system	IC_{50} (μg/ml)	IC_{50} (μg/ml)	IC_{50} (μg/ml)	IC_{50} (μmol/ml)
NA	5.0	0.3	0.04	0.1
5HT	2.0	0.3	0.11	0.2
DA	1.0	0.1	0.06	0.1
GABA	1.0	0.1	0.10	0.2
L-Glutamate	21.0	3.0	0.45	0.8

Half maximal inhibitory concentrations were obtained from [9].

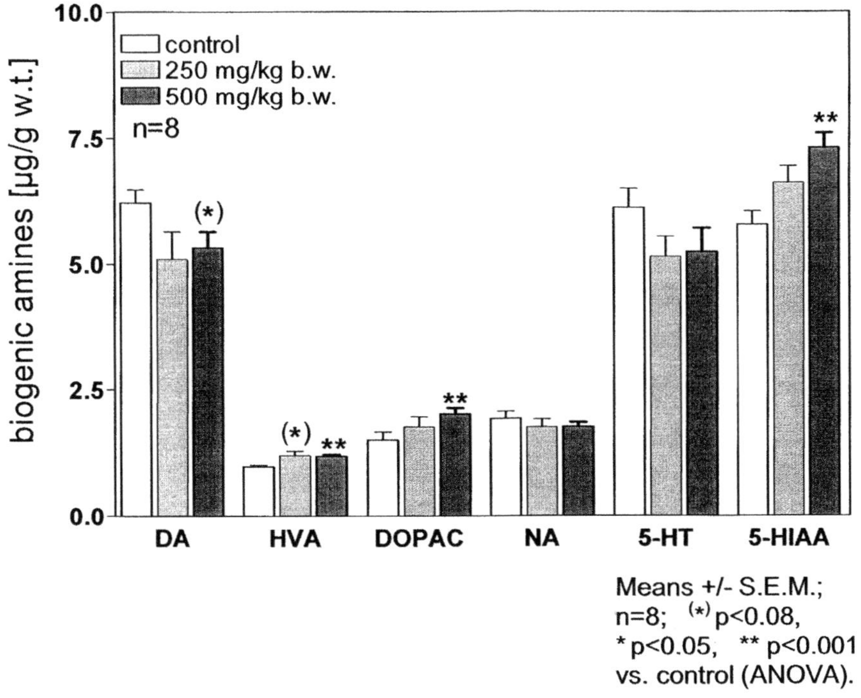

Figure 4. Effects of a single oral administration of two different doses of *Hypericum* extract (hyperforin content 5%) on the levels of serotonin, norepinephrine, and dopamine and of some of their metabolites in the mouse brain. Data are from [12].

subchronic treatment [7, 16, 17]. Our original findings about classical β-down-regulation using the rather non-selective ligand dihydoralprenolol [7] were recently confirmed using a more specific ligand (Fig. 5). Moreover, subchronic treatment with St. John's Wort extract also alters the densities of serotonin and dopamine transporters in rat brain [17].

Hyperforin and adhyperforin are the major uptake inhibiting constituents of St. John's Wort

The MAO-A inhibitory effect of St. John's Wort extract [1] which, however, was subsequently never confirmed, was originally associated with the naphthodianthrone derivative hypericin. Hypericin was for a long time considered the main active constituent of St. John's Wort extract, possibly because it is responsible for the phototoxic properties of the extract [18]. However, MAO inhibition could never be confirmed for hypericin [2–7]. Moreover, hypericin alone was not active in inhibiting the neuronal uptake of serotonin, norepinephrine, dopamine, GABA or L-glutamate [19, 20]. To our surprise, hyper-

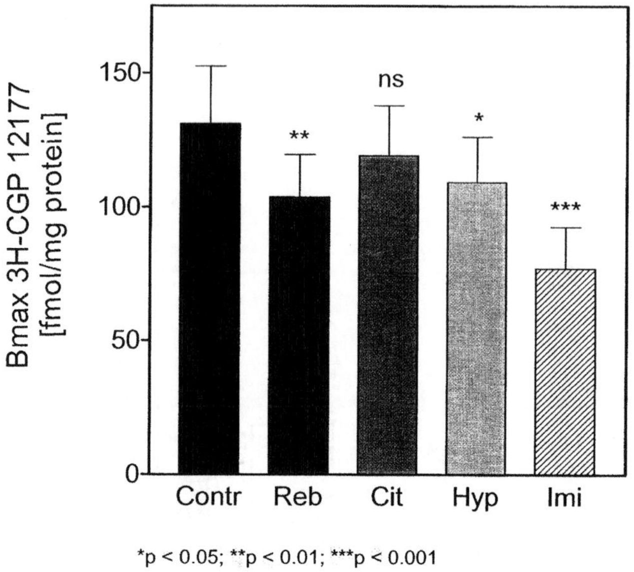

Figure 5. Effect of subchronic treatment (14 days) with Reboxetin, Citalopram, or Imipramine (2×10 mg/kg daily p.o.) or with *Hypericum* extract (2×150 mg/kg daily p.o.) on β receptor density in the frontal cortex of rats as determined by specific ^{3}H-CGP 12177 binding. Data are from [51].

forin, a phloroglucinol derivative, considered for years as not relevant for the biological activity, was soon identified as the constituent mainly responsible for the inhibition of neurotransmitter uptake, since uptake inhibition correlated with hyperforin content (Tab. 2) [7, 9]. However, even these early original findings showed a weak reuptake inhibition that could not been explained by hyperforin [7, 9]. Hyperforin is more than 10 times more potent compared to hypericum extract in neuronal tissues and even more potent in other cell systems [7, 8], which is plausible as hyperforin is quantitatively the most important ingredient in St. John's Wort extract (2–5%). Wonnemann et al. (2001) investigated all relevant single constituents of St. John's Wort extract as possible inhibitors of the synaptosomal uptake of serotonin, norepinephrine, and L-glutamate (Tab. 3). Except hyperforin and adhyperforin, the only fraction showing uptake inhibition were the polyphenols (oligomeric procyanidin fraction) which showed IC_{50} values between 10–30 μg/ml (Fig. 6). Since these IC_{50} values were higher than the values found for St. John's Wort extract containing hyperforin (around 1 μg/ml), we originally considered this effect as a rather unspecific inhibition of transporter protein function. However, since recent findings indicate activity of several flavonoids in behavioural paradigms typical for antidepressant activity (see Chapter by M. Nöldner), and since hyperforin free extracts have been shown to alter brain concentrations of several neurotransmitters [21], it could be possible that the uptake inhibition by polyphenols *in vitro* is much more relevant than originally thought.

Table 3. Synaptosomal uptake inhibition by most relevant constituents of St. John's Wort

	IC_{50} values (μM)		
Substance	5HT uptake	NE uptake	L-glu uptake
Hypericin	> 100	> 10	> 10
Hypericin/pseudohypericin	> 10	> 10	> 10
Kaempferol	> 100	> 10	> 10
Hyperoside	> 100	> 100	> 100
Biapigenin	> 100	> 10	> 10
Quercitrin	> 100	> 100	> 100
Isoquercitrin	> 100	> 10	> 10
Rutin	> 100	> 100	> 100
Armentoflavone	> 100	> 100	> 100
Myricetin	> 10	> 10	> 10
Catechines	n.d.	> 100	> 100
Adhyperforin	0.32 ± 0.10	0.67 ± 0.10	2.40 ± 0.78
Hyperforin	0.205 ± 0.045	0.08	0.14 ± 0.09

Half-maximal inhibitory concentrations were taken from [23].

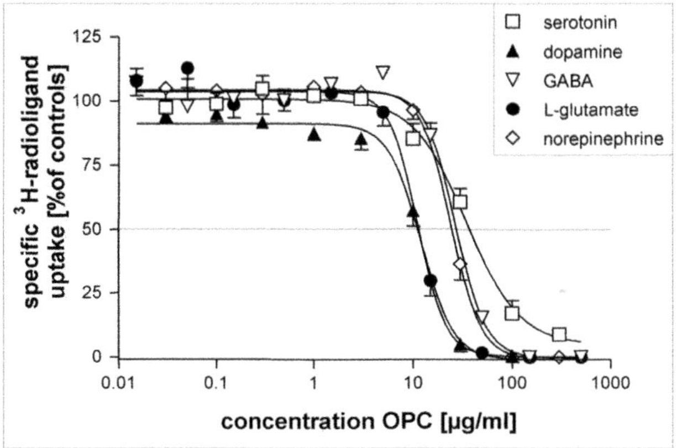

IC_{50} [µg/ml]	
5-HT	34.94 ± 5.5
NE	26.25 ± 2.7
DA	11.71 ± 2.3
GABA	27.65 ± 3.9
L-glu	11.72 ± 3.7

Figure 6. Inhibition of synaptosomal uptake of five neurotransmitters by the oligomeric procyanidin fraction of *Hypericum* extract. Data are from [23].

Hyperforin's and adhyperforin's uptake inhibition is non-competitive

In contrast to all other antidepressants like amitriptyline or citalopram, hyperforin and adhyperforin are non-competitive inhibitors of the monoamine transporter proteins. The kinetic analyses of the inhibition of serotonin, GABA and

L-glutamate uptake by hyperforin show that hyperforin decreases V_{max} but does not increase K_m [22, 23] (see also Tab. 4). Moreover, in contrast to well known antidepressants, hyperforin's inhibitory effect on ^{3}H-paroxetine binding is much weaker than its effect on ^{3}H-serotonin uptake in rat brain synaptosomes (Tab. 5). Binding of ^{3}H-paroxetine to brain membranes labels the serotonin binding site of the human or rodent serotonin transporter molecules. Similar observations were obtained by Gobbi et al. [10] also using ^{3}H-citalopram binding inhibition. Moreover, neither hyperforin nor adhyperforin inhibit the binding of the cocaine analogue, [^{3}H]WIN 35,428 [11] to the dopamine transporter at dopamine uptake inhibiting concentrations. Taken together, these findings indicate that hyperforin does not inhibit neurotransmitter uptake via direct interaction with the specific binding sites of the neurotransmitter transporter molecules.

Table 4. Kinetic analysis of the inhibition of serotonin uptake into mouse brain synaptosomes by citalopram and hyperforin

	K_m [nM]	V_{max} [pmol/min/mg]
Control	12.72 ± 5.01	0.105 ± 0.044
Citalopram	29.4 ± 5.25***	0.106 ± 0.074
Control	11.90 ± 2.07	0.191 ± 0.034
Hyperforin	9.93 ± 0.59*	0.083 ± 0.032***

*p < 0,05; ***p < 0,001. Data are taken from [22].

Table 5. Half-maximal inhibitory concentrations (IC_{50}) for specific ^{3}H-serotonin uptake, and for specific ^{3}H-paroxetine binding. Data are means ± SD.

Substance	^{3}H-paroxetine binding	^{3}H-serotonin uptake	Ratio
	IC_{50} (nM)		
Sertraline	3.3 ± 0.4	2.1 ± 1.0	1.9
Citalopram	4.1 ± 0.6	1.1 ± 0.2	3.7
Fuvoxamine	13.5 ± 3.7	10.9 ± 5.6	1.2
Fluoxetine	26.3 ± 0.6	10.4 ± 5.1	2.6
Desipramine	672 ± 172	214 ± 90	3.1
Hyperforin	19528 ± 6983	1357 ± 171	14.4

Data are taken from [22].

But what is the mechanism of hyperforin's inhibition of the neurotransmitter transport? The monoamine transporters are a subfamily within the large superfamily of Na^+/Cl-dependent neurotransmitter transporters, which also includes the GABA transporter. This superfamily of transporters is characterised by direct coupling of substrate entry to an inward cotransport of sodium ions, which provides the energetic driving force for substrate accumulation within the cell. As Cl is also required for the transporter activity, the family of neurotransmitter transporters is referred to as the Na^+/Cl-dependent transporter family [24]. The transporters for L-glutamate, however, are not structurally related to the above mentioned superfamily, but substrate transport is also coupled directly to cotransport of Na^+ ions [25]. If the intracellular sodium concentration ($[Na^+]_i$) is elevated or the extracellular sodium reduced, the driving force for substrate accumulation is lost.

An additional alternative to influence the synaptosomal uptake is an interference with the storage of monoamines in synaptic vesicles. In intact neurons, a decrease in vesicular storage capacity of monoamines is expected to increase their cytoplasmic concentrations. This increase in cytoplasmic concentrations of monoamines could in turn decrease the transmembrane gradient of neurotransmitters. Consequently, an apparent inhibition of synaptosomal uptake is observed. This mechanism is also relevant for the storage of neurotransmitters in vesicles.

Hyperforin elevates $[Na^+]_i$

In agreement with the first possible mechanism, our studies using ion-specific fluorescence dyes clearly indicate that hyperforin elevates $[Na^+]_i$ in human platelets and PC12 cells [22, 26] at concentrations which are also required for uptake inhibition (Fig. 7). To confirm a causal association between $[Na^+]_i$ and uptake hyperforin was compared with the sodium ionophore monensin, which increases $[Na^+]_i$ non-specifically by generating membrane pores. Both compounds elevated $[Na^+]_i$ over basal levels in human platelets at the same concentration needed to inhibit serotonin uptake (Fig. 8). Furthermore, monensin also inhibits the uptake of noradrenaline, dopamine, GABA and L-glutamate with similar IC_{50} values as hyperforin does (unpublished results from our laboratory). Again, like hyperforin, monensin also represents a non-competitive serotonin uptake inhibitor. These findings affirm the relevance of the reduced sodium gradient for the uptake inhibition.

Even though hyperforin resembles in many ways the sodium ionophore monensin, our subsequent data indicate that hyperforin is not a sodium ionophore. In contrast to monensin, hyperforin shows a reverse U-shaped dose-response curve and does not elevate intracellular sodium to the extracellular level like monensin [22]. These results allow the hypothesis that hyperforin activates a specific ion conductive mechanism. This could explain that its effects on intracellular sodium are determined once a certain level of $[Na^+]_i$ is reached. The findings that hyperforin leads to an elevation of $[Na^+]_i$ not only explain easily its effects on serotonin uptake in platelets and synaptosomes, but also explain its non-selective profile of uptake inhibition on many neurotransmitter transport systems as well as for choline [27].

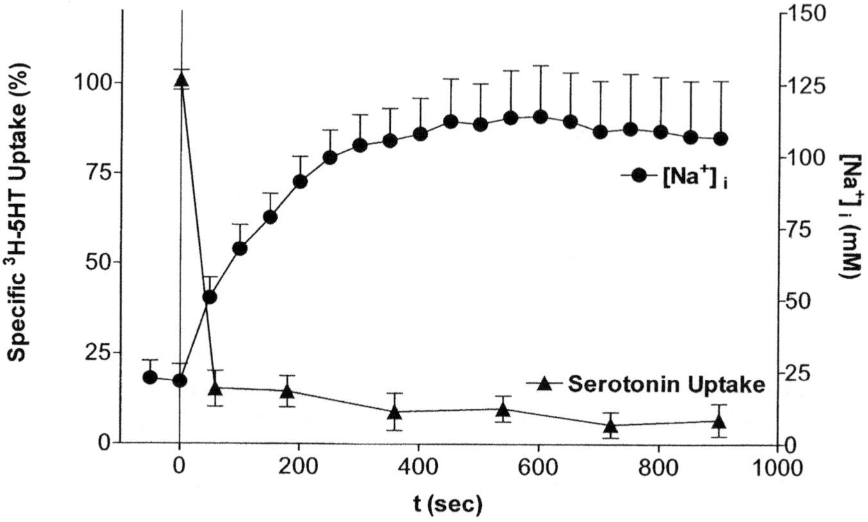

Figure 7. Time dependency of hyperforin's (10 μM) effects on $[Na^+]_i$ and serotonin uptake in human platelets. Specific serotonin uptake (filled circle) was measured first after one min, afterwards every 30 s over 15 min. $[Na^+]_i$ was also followed for 15 min. Already after 1 min, hyperforin shows nearly maximum inhibition of specific serotonin uptake. This effect is stable over 15 min. Additionally, hyperforin elevates $[Na^+]_i$ from 20 mM to about 100 mM (filled triangle). After 5 min, a plateau is reached. Data are adapted from [22].

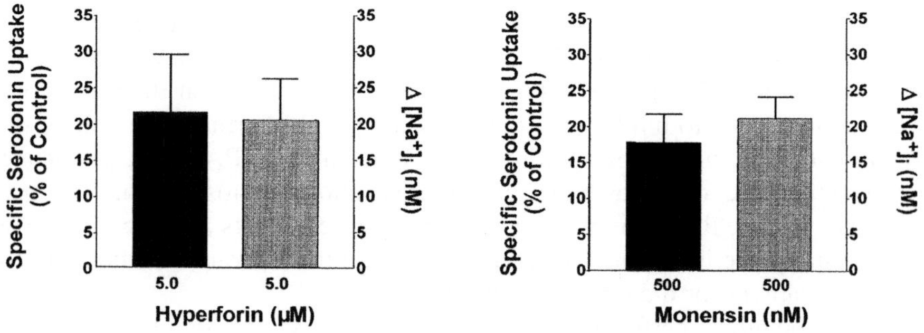

Figure 8. Hyperforin and the sodium ionophore monensin elevate $[Na^+]_i$ in human platelets both by about 20 mmol/L at concentrations that inhibit ^{3}H-serotonin uptake by about 80%. Data are from [22].

Hyperforin activates non-selective cation channels

Different sodium conductive pathways play an important role in regulating the $[Na^+]_i$ in synaptosomes, human platelets, and PC12 cells: the Na^+-K^+ ATPase, voltage-dependent sodium channels, the Na^+/H^+ exchanger, the Na^+/Ca^{2+}-

exchanger and non-selective cation channels (NSCCs). Our extensive research did not indicate that hyperforin is either an inhibitor of Na^+-K^+ ATPase, voltage-dependent sodium channels, the Na^+/H^+ exchanger or the Na^+/Ca^{2+}-exhanger [22, 26, 28]. Therefore, we assumed that NSCCs could play a role in the observed elevation of $[Na^+]_i$ by hyperforin. These channels are permeable for mono- and divalent cations, especially for sodium and calcium ions. Our subsequent finding that hyperforin additionally elevates concentration dependently $[Ca^{2+}]_i$ in different cell types further supports this hypothesis as shown for its effects for $[Ca^{2+}]_i$ in PC12 cells (Fig. 9).

NSCCs were identified in platelets [29], synaptosomes [30] and PC12 cells [31]. They can be inhibited by a variety of inhibitors: SK&F 96365, LOE 908, flufenamic acid and the two lanthanides Gadolinium and Lanthan [32–37]. Since all these inhibitors affect the hyperforin induced sodium and calcium elevation ([26]; see also Fig. 9), it is quite plausible that hyperforin's inhibition of neurotransmitter uptake can be explained by NSCCs activation.

Part of the NSCCs belongs to the family of the TRP channels, for which many different genes were identified [38]. A possible overlap between NSCCs and several TRP channels has been proposed on the basis of some common pharmacological properties [39]. While the pharmacological investigation of TRP channels is currently under investigation, the high sensitivity of some TRP channels for the lanthanides Gd^{3+} and La^{3+} at micromolar concentrations seems to be a rather specific property [37]. Very importantly, both lanthanides

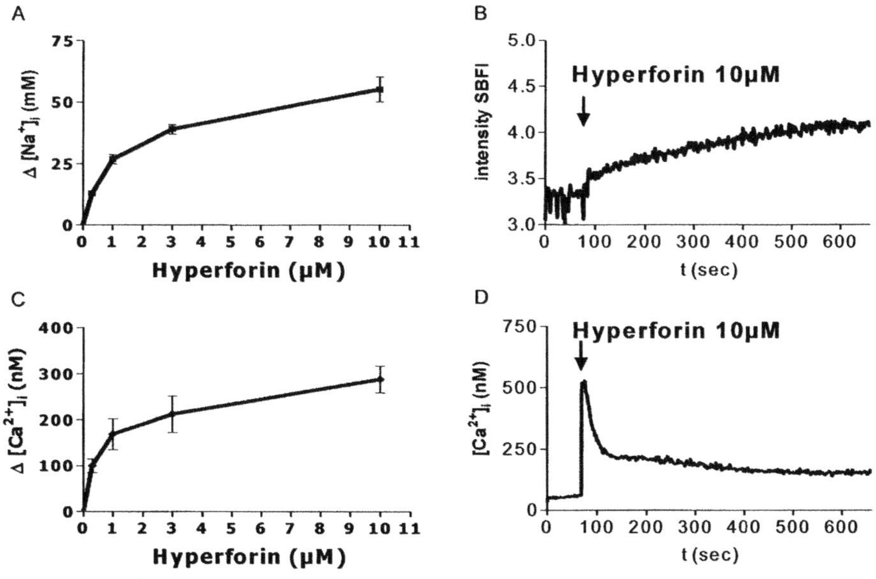

Figure 9. Hyperforin elevates concentration dependently $[Ca^{2+}]_i$ (C) and $[Na^+]_i$ (A) in PC12 cells. Time dependency for the elevation of $[Ca^{2+}]_i$ and $[Na^+]_i$ by hyperforin 10 µM is shown in D and B. Data are from [26].

inhibit the hyperforin-induced sodium and calcium increase in PC12 cells and attenuated the hyperforin-induced inhibition of specific serotonin-uptake into human platelets [26]. Summarising, our results indicate an activation of NSCCs or TRP channels by hyperforin. This activation leads to an elevated intracellular sodium and calcium concentration and results in the inhibition of ^{3}H-serotonin uptake in human platelets. Our novel findings that hyperforin finally modulates several neurotransmitter systems in the brain by activating NSCCs or TRP channels make it the first member of a new possible class of antidepressant drugs.

Hyperforin modulates membrane fluidity

Although rather good evidence indicates the association of hyperforin's mechanism of action with NSCCs, the molecular mechanism by which hyperforin activates NSCCs remains uncertain. NSCCs are activated by rather heterogeneous stimuli. They are mainly activated by G-protein-coupled receptors or tyrosine kinase coupled receptors and to some extent by intracellular calcium ions, but also by hydrostatic pressure or stretching, ATP and further unknown parameters [40–42]. Very interestingly, hyperforin modifies brain membrane fluidity *in vitro* at rather low concentrations [43]. Considering the importance of the physiochemical state of the membrane for G-protein activity, changes in membrane fluidity could initiate activation of NSCCs or TRP channels. GTP-binding proteins are membrane associated proteins that underlie conformational changes, including membrane association and dissociation of the alpha subunit during the GTP cycle stimulated by agonist-bound receptors. Membrane fluidity is depending on cholesterol concentration. Accordingly, it was shown that cholesterol modulates GTPase activity of G Proteins in the cell membrane [44]. Furthermore, membrane flexibility and phospholipid composition modulate GTPase activity of G proteins [45]. Since our initial findings were related to *in vitro* experiments, we have recently investigated brain levels of hyperforin and membrane changes after acute oral treatment of mice with pharmacological relevant doses of St. John's Wort extract and hyperforin [46, 47]. In hyperforin or St. John's Wort extract treated mice, hyperforin concentrations of 53.7 ± 18.7 pmol/g or 29.5 ± 20.3 pmol/g, respectively, were found. Both treatments lead to decreased annular- and bulk fluidity and increased acyl-chain flexibility of brain membranes. Hyperforin-related fluidity changes were significantly correlated with hyperforin brain levels. In the St. John's Wort extract treated animals, the effects were less pronounced in agreement with lower brain levels (Fig. 10). Our data emphasise a membrane interaction of hyperforin that possibly contributes to its effects on ion transduction pathways.

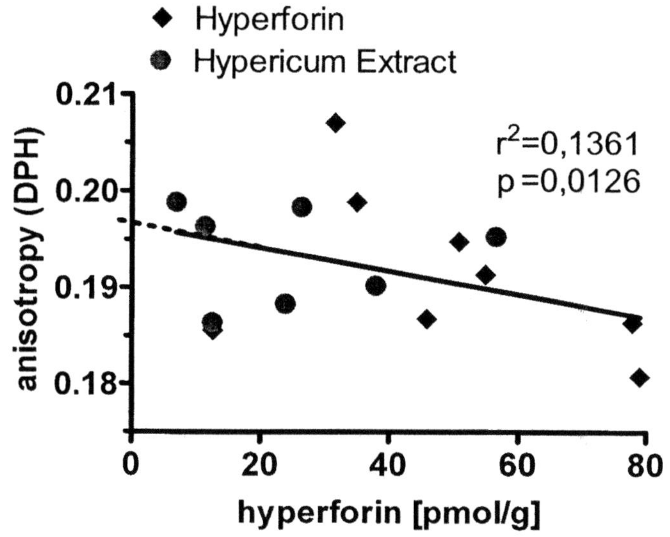

Figure 10. Correlation between hyperforin brain levels and changes of membrane fluidity (DPH anisotropy) in mice treated with hyperforin 15 mg/kg or with *Hypericum* extract 300 mg/kg containing 5% hyperforin. Data are from [47].

Hyperforin inhibits vesicular uptake of monoamines and induces efflux of monoamines

As mentioned before, an additional possibility of inhibiting synaptosomal uptake would be inhibition of vesicular uptake of monoamines and the subsequent release of monoamines from vesicles. Recently Roz et al. demonstrated that hyperforin inhibits dose dependently and equipotently (in comparison to synaptosomal uptake inhibition) the uptake of serotonin, dopamine and noradrenaline in a non-competitive manner into rat brain synaptic vesicles. Furthermore, hyperforin induced an efflux of serotonin in preloaded synaptic vesicles. It was shown that hyperforin does not interact with vesicular monoamine transporters. Roz et al. [48] discuss that hyperforin acts similar to the protonophore FCCP and interferes with the driving force of vesicular uptake, the pH gradient across the vesicular membrane. They propose that hyperforin, like the protonophore FCCP, dissipates an existing pH gradient by an efflux of inwardly pumped protons. This hypothesis is supported by our findings that hyperforin lowers intracellular pH transiently [49, 50] and the findings of Gobbi et al. that Ro 04-1284, a reserpine-like compound, potently reduces synaptosomal ^{3}H-serotonin accumulation, although this effect was not complete (maximal inhibition 60%). However, the compound additionally does inhibit ^{3}H-citalopram binding to the serotonin transporter. Both mechanisms would not explain the inhibitory effect of hyperforin on GABA and L-glutamate transport.

Conclusion

The cumulated data so far available suggests the following scenario to explain the preclinical antidepressant properties of hyperforin and adhyperforin which are rather unselective inhibitors of serotonin, dopamine, noradrenaline, GABA, and L-glutamate uptake. Although the detailed molecular mechanism of uptake inhibition is still under investigation, most evidence points to an elevation of $[Na^+]_I$, which in turn leads to an inhibition of the sodium dependent neurotransmitter transporters. The increase in $[Na^+]_i$ and additionally $[Ca^{2+}]_i$ is due to the activation of NSCCs or TRP channels, possibly associated with changed membrane fluidity. Additionally, hyperforin inhibits vesicular uptake of monoamines and release of monoamines from vesicles. Taken together, this mode of action is speculatively followed by reduced presynaptic uptake and an increased release of several neurotransmitters and to changes of their extracellular as well as intracellular concentrations *in vivo*. Hyperforin's broad spectrum of pharmacological activity might explain pharmacological effects beyond those of typical antidepressant drugs. Hyperforin's beneficial effects on cognition and its modulation of the processing of the β-amyloid precursor protein might be examples [50, 51].

References

1 Suzuki O, Katsumata Y, Oya M, Bladt S, Wagner H (1984) Inhibition of monoamine oxidase by hypericin. *Planta Med* 50: 272–274
2 Bladt S, Wagner H (1994) Inhibition of MAO by fractions and constituents of *Hypericum* extract. *J Geriatr Psychiatry Neurol* 7 (1): S57–59
3 Thiede HM, Walper A (1994) Inhibition of MAO and COMT by *Hypericum* extracts and hypericin. *J Geriatr Psychiatry Neurol* 7 (1): S54–S56
4 Reuter HD (1993) *Hypericum* als pflanzliches Antidepressivum. *Zeitschrift für Phytotherapie* 14: 239–254
5 Demisch L, Hölzl J, Gollnik B, Kaczmarczyk P (1989) Identification of selective MAO-Type A inhibitors in *Hypericum perforatum* L. (Hyperforat). *Pharmacopsychiatry* 22: 194
6 Cott JM (1997) *In vitro* receptor binding and enzyme inhibition by *Hypericum perforatum* extract. *Pharmacopsychiatry* 30: 108–112
7 Müller WE, Rolli M, Schäfer C, Hafner U (1997) Effects of *Hypericum* extract (LI 160) in biochemical models of antidepressant activity. *Pharmacopsychiatry* 30: 102–107
8 Neary JT, Whittemore SR, Bu Y, Mehta H, Shi YF (2001) Biochemical mechanisms of action of *Hypericum* LI160 in glial and neuronal cells: inhibition of neurotransmitter uptake and stimulation of extracellular signal regulated protein kinase. *Pharmacopsychiatry* 34: S103–107
9 Chatterjee SS, Bhattacharya SK, Wonnemann M, Singer A, Müller WE (1998) Hyperforin as a possible antidepressant component of *Hypericum* extracts. *Life Sci* 63: 499–510
10 Gobbi M, Valle FD, Ciapparelli C, Diomede L, Morazzoni P, Verotta L, Caccia S, Cervo L, Mennini T (1999) *Hypericum perforatum* L. extract does not inhibit 5-HT transporter in rat brain cortex. *Naunyn Schmiedebergs Arch Pharmacol* 360: 262–269
11 Jensen AG, Hansen SH, Nielsen EO (2001) Adhyperforin as a contributor to the effects of *Hypericum perforatum* L. in biochemical models of antidepressant activity. *Life Sci* 68: 1593–1605
12 Serdarevic N, Eckert GP, Müller WE (2001) The effects of extracts from St. John's Wort and Kava Kava on brain neurotransmitter levels in the mouse. *Pharmacopsychiatry* 34: 134–136
13 Rommelspacher H, Siemanowitz B, Mannel M (2001) Acute and chronic actions of a dry

methanolic extract of *Hypericum perforatum* and a hyperforin-rich extract on dopaminergic and serotoninergic neurones in rat nucleus accumbens. *Pharmacopsychiatry* 34: 119–126
14 Phillipu A (2001) *In vivo* neurotransmitter release in the loceus coereleus – effects of hyperforin, inescapable shock and fear. *Pharmacopsychiatry* 34: S111–115
15 Calapai G, Crupi A, Firenzuali F, Inferrera G, Squadrito F, Parisi A (2001) Serotonin, norepinephrine and dopamine involvement in the antidepressant action of *Hypericum perforatum*. *Pharmacopsychiatry* 34: 45–49
16 Teufel-Mayer R, Gleitz J (1997) Effects of long-term administration of *Hypericum* extracts on the affinity and density of central serotonergic 5HT-1 A and 5HT-2 A receptors. *Pharmacopsychiatry* 30 (2): 113–116
17 Chen F, Rezvani AH, Lawrence AJ (2003) Autoradiographic quantification of neurochemical markers of serotonin, dopamine and opioid systems in rat brain mesolimbic regions following chronic St. John's Wort treatment. *Naunyn-Schmiedeberg's Arch Pharmacol* 367: 126–133
18 Brockmoller J, Reum T, Bauer S, Kerb R, Hubner WD, Roots I (1997) Hypericin and pseudohypericin: pharmacokinetics and effects on photosensitivity in humans. *Pharmacopsychiatry* 30 (Suppl 2): 94–101
19 Müller WE, Schäfer C (1996) Johanniskraut. *In vitro*-Studie über *Hypericum* Extrakt, Hypericin und Kämperol als Antidepressiva. *Dtsch Apoth Ztg* 136: 1015–1022
20 Rolli M, Schäfer C, Müller WE (1995) Effect of *Hypericum* extract on neurotransmitter receptor binding and synaptosomal uptake systems. *Pharmacopsychiatry* 28: 207
21 Misane I, Ogren SO (2001) Effects of *Hypericum perforatum* (St. John's Wort) on passive avoidance in the rat: evaluation of potential neurochemical mechanisms underlying its antidepressant activity. *Pharmacopsychiatry* 34 (Suppl 1): S89–97
22 Singer A, Wonnemann M, Müller WE (1999) Hyperforin, a major antidepressant constituent of St. John's Wort, inhibits serotonin uptake by elevating free intracellular Na^+. *J Pharmacol Exp Ther* 290: 1363–1368
23 Wonnemann M, Singer A, Siebert B, Muller WE (2001) Evaluation of synaptosomal uptake inhibition of the most relevant constituents of St. John's Wort. *Pharmacopsychiatry* 34: S148–152
24 Giros B, Caron MG (1993) Molecular characterization of the dopamine transporter. *Trends Pharmacol Sci* 14: 43–49
25 Kanner BI, Danbolt N, Pines G, Koepsell H, Seeberg E, Mathisen JS (1993) Structure and function of the sodium and potassium-coupled glutamate transporter from rat brain. *Biochem Soc Trans* 21: 59–61
26 Treiber K, Singer A, Henke B, Müller WE (2004) Hyperforin is a specific activator of non-selective cation channels. *Brit J Pharmacol; in press*
27 Buchholzer ML, Dovrak C, Chatterjee S, Klein J (2002) Dual modulation of striatal acetylcholine release by hyperforin, a constituent of St. John's Wort. *JPET* 301: 1–6
28 Wonnemann M, Singer A, Müller WE (2000) Inhibition of synaptosomal uptake of ^{3}H-L-glutamate and ^{3}H-GABA by hyperforin, a major constituent of St. John's Wort. The role of amiloride sensitive sodium conductive pathways. *Neuropsychopharmacology* 23: 188–197
29 Stamouli V, Vakirtzi-Lemonias C, Siffert W (1993) Thrombin and NaF, but not epinephrine, raise cytosolic free Na^+ in human platelets. *Biochim Biophys Acta* 1176: 215–221
30 Reichardt LF, Kelly RB (1983) A molecular description of nerve terminal function. *Annu Rev Biochem* 52: 871–926
31 Tesfai Y, Brereton HM, Barritt GJ (2001) A diacylglycerol-activated Ca^{2+} channel in PC12 cells (an adrenal chromaffin cell line) correlates with the expression on TRP-6 (transient receptor potential) protein. *Biochem J* 358: 717–726
32 Merritt JE, Armstrong WP, Benham CD (1990) SKandF 96365, a novel inhibitor of receptor-mediated calcium entry. *Biochem J* 271: 515–552
33 Goegelein H, Dahlem D, Englert HC, Lang HJ (1990) Flufenamic acid, mefenamic acid and niflumic acid inhibit single non-selective cation channels in the rat exocrine pancreas. *FEBS Lett* 268: 79–82
34 Kawanabe Y, Hashimoto N, Masaki T (2002) Characterisation of G proteins involved in activation of nonselective cation channels by endothelin (B) receptor. *Br J Pharmacol* 136: 1015–1022
35 Krautwurst D, Seifert R, Hescheler J, Schultz G (1992) Formyl peptides and ATP stimulate Ca^{2+} and Na^+ inward currents through non-selective cation channels via G-proteins in dibutyryl cyclic AMP- differentiated HL-60 cells. Involvement of Ca^{2+} and Na^+ in the activation of beta-glucuronidase release and superoxide production. *Biochem J* 288 (Pt 3): 1025–1035

36 Miwa S, Iwamuro Y, Zhang XF, Kawanabe Y, Masaki T (2000) LOE 908: a specific blocker of non-selective cation channel. *Cardiovasc Drug Rev* 18: 61–72
37 Jung S, Mühle A, Schaefer M, Strotmann R, Schultz G, Plant TD (2003) Lanthanides potentiate TRPC5 currents by an action at extracellular sites close to the pore mouth. *J Biol Chem* 278 (6): 3562–3571
38 Clapham DE (2003) TRP channels as cellular sensors. *Nature* 426: 517–524
39 Inoue R, Okada T, Onoue Y, Hara Y, Shimizu S, Naitoh S, Ito Y, Mori Y (2001) The transient receptor potential protein homologue TRP6 is the essential component of vascular α1-adrenoceptor-activated Ca^{2+}-permeable cation channel. *Circ Res* 88: 325–332
40 Suzuki M, Murata M, Ikeda M, Miyoshi T, Imai M (1998) Primary structure and functional expression of a novel non-selective cation channel. *Biochem Biophys Res Commun* 242: 191–196
41 Liu RJ, van den Pol AN, Aghajanian GK (2002) Hypocretins (Orexin) regulate serotonin neurons in the dorsal raphe nucleus by excitatory direct and inhibitory indirect actions. *J Neurosci* 22: 9453–9464
42 Clapham DE (2003) TRP channels as cellular sensors. *Nature* 426: 517–524
43 Eckert GP, Müller WE (2001) Effects of hyperforin on the fluidity of brain membranes. *Pharmacopsychiatry* 34 (Suppl 1): S22–25
44 Wood EG, Eckert GP, Igbavboa U, Müller WE (2003) Amyloid beta-protein interactions with membranes and cholesterol: causes or causality for Alzheimer's disease. *Biochim Biophys Acta* 1610: 281–290
45 Ropero S, Chiloeches A, Montes A, Toro-Nozal MJ (2003) Cholesterol cell content modulates GTPase activity of G-proteins in GH4C1 cell membranes. *Cell Signall* 15: 131–138
46 Keller JH, Karas M, Müller WE, Volmer DA, Eckert GP, Tawab MA, Blume HH, Dingermann T, Schubert-Zsilavecz M (2003) Determination of hyperforin in mouse brain by high-performance liquid chromatography/tandem mass spectrometry. *Anal Chem* 75: 6084–6088
47 Eckert GP, Keller JH, Jourdan C, Karras M, Volmer DA, Schubert-Zsilavecz M, Müller WE (2004) Hyperforin modifies neuronal membrane properties *in vivo*. *Neurosci Lett* 367: 139–143
48 Roz N, Rehavi M (2003) Hyperforin inhibits vesicular uptake of monoamines by dissipating pH gradient across synaptic vesicle membrane. *Life Sci* 73: 461–470
49 Froestl B, Steiner B, Müller WE (2003) Enhancement of proteolytic processing of the β-amyloid precursor protein by hyperforin. *Biochem Pharmacol* 66: 2177–2184
50 Klusa V, Germane S, Noldner M, Chatterjee SS (2001) *Hypericum* extract and hyperforin: memory-enhancing properties in rodents. *Pharmacopsychiatry* 34: S61–69
51 Holoubek G, Nöldner M, Müller WE (2001) Possible relationship between behavioral and biochemical changes following antidepressant activity. *Soc Neurosci*, Abstr. 572.14

St. John's Wort and its Active Principles in Depression and Anxiety
Edited by W. E. Müller

Modulation of neurotransmitter release and metabolism

Jan Kehr, Sven Ove Ögren and Takashi Yoshitake

Department of Neuroscience, Karolinska Institutet, Retzius väg 8, 171 77 Stockholm, Sweden

St. John's Wort and the monoamine hypothesis of depression

Extracts from the herb *Hypericum perforatum* L. (St. John's Wort) have become increasingly used for the treatment of mild-to-moderate depression [1], most likely as a consequence of their good tolerability and prescription-free accessibility (for review, see [2, 3]). A number of constituents from the extracts of St. John's Wort have recently been identified and tested for their biological activity (for review, see [4]). Most studies on antidepressant efficacy and mechanisms of action of St. John's Wort have searched for biochemical and behavioural evidence linking the *Hypericum* extract or its components to the monoamine hypothesis of depression. This hypothesis postulates brain monoaminergic hypofunction as the major neurochemical factor behind depression [5]. It is based mainly on clinical evidence on therapeutic efficacy of antidepressant drugs such as monoamine oxidase inhibitors (MAOI), tricyclic antidepressants (TCAs), selective serotonin (SSRI) or noradrenaline reuptake inhibitors or mixed 5-HT/NA reuptake inhibitors (NSRI). All these drugs share the ability to increase brain synaptic levels of serotonin (5-HT), noradrenaline (NA) and/or dopamine (DA) or all three monoamines and thereby increase monoaminergic neurotransmission.

Most biochemical studies on mechanisms of St. John's Wort extracts have addressed the question whether the extracts or their individual components resemble the action of synthetic antidepressant drugs on monoamine transmission. A number of behavioural studies were also carried out to confirm the potency of St. John's Wort extracts in animal models of depression. Several *in vitro* studies on synaptosomal and vesicular reuptake have provided experimental evidence for the view that hydro-alcoholic extracts of *Hypericum perforatum* and particularly one of its principal components hyperforin, can facilitate monoaminergic (5-HT, NA, DA) and amino acid (glutamate, GABA) transmission (for review, see [3]). However, an increasing body of evidence suggests that the mechanism of action of *Hypericum* and hyperforin is far more complex than that ascribed to SSRIs (for review, see [2–4]). Besides hyperforin, other active components or metabolites have been implicated in the overall activity of the plant extracts. These components are believed to exert synergetic effects, which can only be detected by *in vivo* studies following chronic administration of St. John's Wort extracts. The *in vivo* approach to study brain neurotransmission is well established and allows investigations of double- and triple-acting monoamine reuptake inhibitors (for review, see [6]). A number of studies have documented marked differences between the mechanisms of action of central nervous system (CNS) drugs when comparing their *in vitro* and *in vivo* profiles. For instance, several microdialysis studies have shown that NSRIs such as venlafaxine [7] and duloxetine [8] differ in their potency of inhibiting the reuptake of 5-HT and NA when comparing results obtained *in vitro* and *in vivo*.

The aim of this Chapter is to review the actions of St. John's Wort extracts on neurotransmitter release and metabolism by analysing the results obtained by *ex vivo* methods and *in vivo* techniques such as microdialysis.

Major neurotransmitters, their receptors and transporter proteins implicated in the action of St. John's Wort

A frequently emphasised pharmacological distinction of St. John's Wort as compared to conventional antidepressants is its proposed ability to modulate multiple neurotransmitter systems [3, 4]. Besides the monoamines (5-HT, NA and DA), *Hypericum* extracts have been shown to modulate release of glutamate (Glu), GABA, and acetylcholine (ACh). *Hypericum* or its constituents also exert relatively high binding affinity to several receptor subgroups [4] in brain areas, which are implicated in pathophysiology of mood disorders. This fact provides a rationale to further explore the mechanisms of action of *Hypericum* extract and its constituents *in vivo*, i.e., under conditions of intact brain circuitry and integrated physiological functions.

Noradrenaline

Noradrenergic system in the brain has been targeted by antidepressant drugs since the introduction of tricyclic antidepressants (TCAs) (e.g., amitriptyline) and MAOIs (e.g., phenelzine). Some recently introduced antidepressants also target the NA system, including selective inhibitors of NA reuptake, e.g., reboxetine or α2-adrenoreceptor blockade (mirtazapine). *Hypericum* extract, and particularly its component hyperforin, were shown to inhibit NA uptake into synaptosomes [9, 10] or brain slices [11] at a similar potency as 5-HT and DA. A recent study on *in vitro* binding of pure constituents of St. John's Wort [12] revealed that hyperforin bound to human NET 5–10 times more potently than to human SERT or bovine DAT transfected cells. The latter study also revealed that hypericin potently bound to β-adrenoreceptors and some flavonoid constituents to α2-adrenoreceptors. However, the analysis of tissue content and neurochemical studies *in vivo* provide a weak support for the involvement of the NA system in the action of *Hypericum* extracts. Acute oral administration of *Hypericum* extracts Li 160 and/or Ph-50 at doses of 250 and 500 mg/kg increased brain noradrenaline content only in the diencephalon and brainstem but not in the cortex [13, 14]. No significant changes in NA concentrations were observed in the mouse cortex, hippocampus, hypothalamus and caudate regions following oral administration of *Hypericum* extracts [15]. In addition, Butterweck et al. [16] reported no changes in NA content in rat hypothalamus even after 2 and 8 weeks treatment with *Hypericum* extract (500 mg/kg), whereas chronic hypericin (0.2 mg/kg) caused reduction of the NA levels only in the hippocampus.

The effects of hyperforin on extracellular levels of NA in the locus coeruleus were measured by push–pull superfusion technique in anaesthetised rats [17]. The authors reported a moderate increase (by about 50%) of extracellular NA following administration of a high dose (10 mg/kg i.p.) of hyperforin. In a recent microdialysis study, we have shown that an acute high (60 mg/kg i.p.) but not low (30 mg/kg i.p.) dose of *Hypericum* extract (containing 4.1% of hyperforin) increased extracellular NA levels in the hippocampus but not in the prefrontal cortex (PFC) of awake rats. In a similar manner, a clinically more relevant dose of *Hypericum* given orally (300 mg/kg p.o.) had no effect on the NA efflux in the hippocampus and PFC, which are brain regions implicated in the etiology of depression. The overall effect (expressed as the area under the curve (AUC) during the 0–180 min interval) following a single dose of *Hypericum* extract on the extracellular NA levels measured in the rat hippocampus and PFC is shown in Figure 1. A marginal effect was observed at the highest dose (60 mg/kg i.p.) on the extracellular NA levels in the hippocampus as compared to the potency of the NA reuptake inhibitor desipramine (3 mg/kg i.p.). In addition, in the rat the *Hypericum* extract was found ineffective to modulate pineal melatonin [18], which is released via NA mediated mechanism and may serve as an indirect indicator of β-adrenergic NA function.

In conclusion, the available *in vitro* and *in vivo* data indicate that the noradrenergic system is only marginally involved in the antidepressant properties of *Hypericum* extracts.

Serotonin

The role of 5-HT in antidepressant action of *Hypericum perforatum* extracts has been thoroughly investigated by several groups using both biochemical and behavioural methods (for review, see [2–4]). A current hypothesis on the mechanism of action of *Hypericum* extracts proposes that inhibition of neuronal reuptake of 5-HT by the phloroglucinol hyperforin could serve as the major mechanism behind its antidepressant effect [9, 10, 19–21]. However, several *in vitro* binding and reuptake studies have suggested that the inhibitory effect of

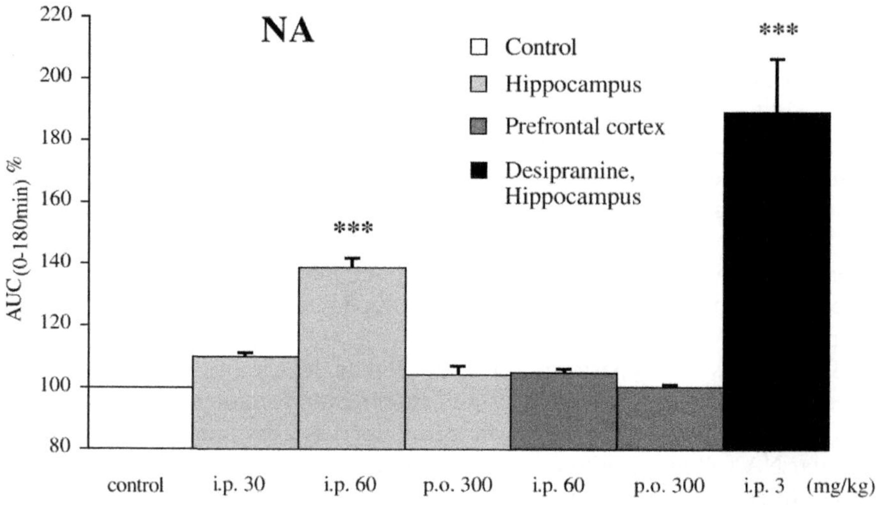

Figure 1. The effect of ethanolic extract of *Hypericum perforatum* on extracellular levels of noradrenaline in the prefrontal cortex and hippocampus of awake rat. *Hypericum* was given to separate groups of rats at 30, 60 kg/kg i.p. and 300 mg/kg p.o. Desipramine (3 mg/kg i.p.) was chosen as a reference compound. The data are expressed as AUCs for time 0–180 min, means ± S.E.M, n = 5 rats, (***) $P < 0.001$, ANOVA, followed by Fisher's PLSD-test. The control group received a 0.2% agar solution used as a solvent to suspend the plant extract. The AUCs for *Hypericum*-treated groups were calculated from the original data on response curves for 20 min fractions of microdialysis samples published elsewhere [29].

Hypericum extracts and hyperforin is mediated via ionic channels and not via a direct action on the SERT protein [20, 22–24]. These *in vitro* experiments argue for a molecular mechanism behind the antidepressant action of *Hypericum* extract, which at least in animal experiments, resembles the action of SSRIs. Some behavioural studies in animal models of depression support this view [25, 26], whereas other investigators suggest a more complex antidepressant profile more reminiscent of TCAs [27]. Thus, *Hypericum* extracts only weekly altered the behavioural effects induced by the 5-HT releasing compound p-chloroamphetamine, which utilises the 5-HT uptake carrier to exert its effects [27]. It was recently shown that both hyperforin and hypericin free extracts, containing high levels of flavonoids, could exert antidepressant activity in the tail suspension and forced swim tests [28]. This finding was supported by biochemical data analysing 5-HT content in the rat cortex, diencephalon and brainstem following acute administration of *Hypericum* extracts with high and low flavonoids content [13, 14]. The dose of the extract (Ph-50) with high (50%) flavonoid content increased 5-HT levels in all brain structures examined, whereas the extract (Li 160) with a low (6%) flavonoid content increased 5-HT content only in the cortex. Measurements of monoamine content in the brain tissue are, as pointed out earlier, difficult to interpret since these values reflect several different and sometimes opposite mechanisms. This is illustrated by the data reporting the increased content of 5-HT in the rat hypothalamus but not in hippocampus, while 5-HIAA levels decreased following a chronic (8 weeks) treatment with *Hypericum* extract (5.3% hyperforin) or hypericin alone [16]. Interestingly, these authors could measure only a minor reduction in 5-HT turnover following 8 weeks treatment with imipramine (15 mg/kg p.o.). Imipramine given for 2 weeks had no effects on 5-HT or 5-HIAA levels in these brain structures. In another study, no effects were detected in 5-HT content in the rat PFC following a single oral dose of *Hypericum* (300 mg/kg) containing 6.7% flavonoids and 4.1% hyperforin [29]. A study in mice [15] revealed that acute *Hypericum* extract caused an increase in 5-HT content in the hypothalamus, hippocampus but not in the cortex and nucleus caudatus, whereas 5-HIAA levels increased in all brain structures. This would suggest that already an acute dose of *Hypericum* may increase 5-HT turnover in agreement with our data [29]. In contrast, antidepressant drugs such as MAOI clorgyline (4 mg/kg i.p.) or phenelzine (5 mg/kg i.p.) were shown to significantly reduce 5-HT turnover in the PFC and striatum of the rat [29, 30].

In summary, the *ex vivo* analysis of the tissue 5-HT and 5-HIAA content following *Hypericum* treatment has provided inconsistent results. This may most likely be explained by the use of different experimental protocols or by dissecting and processing too large (non 5-HT innervated) brain areas. Another possible explanation refers to the discrepancy in pharmacological efficacy of *Hypericum* extracts *in vitro* and *in vivo*. *Hypericum* and its constituent hyperforin compared to some typical antidepressant drugs display a markedly weaker potency to increase 5-HT neurotransmission *in vivo*.

A large body of evidence suggests that in rodents, a single acute dose of an antidepressant drug elicits marked changes in biochemistry and physiology of monoaminergic systems, reflected for example, by increased levels of monoamine neurotransmitters or altered firing activity of the monoaminergic neurons. The microdialysis techniques allows monitoring of extracellular levels of monoamines and metabolites, which provides a unique information on spatial and temporal neurochemistry of neurotransmitters, their release and reuptake at the presynaptic and somato-dendritic levels *in vivo* (for review, see [31]). However, in spite of the acceptance of microdialysis as an optimal technique to study mechanisms of action of antidepressant drugs *in vivo*, there are only a few reports available in which microdialysis was used to examine effects of *Hypericum* extracts on 5-HT release. The microdialysis technique can also be helpful in examining the discrepancies obtained by *in vitro* and *in vivo* methods. However, it is notable that there exists only one microdialysis study supporting the hypothesis that *Hypericum* extract and hyperforin given orally act in a manner similar to the SSRIs [32]. An acute or subchronic administration of methanolic (Li 160, 4.67% hyperforin) or hyperforin-rich (30.14% hyperforin) extracts of *Hypericum perforatum* increased the extracellular concentrations of 5-HT in the nucleus accumbens of awake rats. In another study using the push–pull superfusion technique, systemic administration of hyperforin (10 mg/kg i.p.) caused a moderate increase in 5-HT efflux in the locus coeruleus, similar to that recorded for NA [17]. In a recent microdialysis study [29] and also summarised in Figure 2, the 5-HT concentrations in the PFC or hippocampus of awake rats were only marginally affected by injection of 60 mg/kg i.p. (max. increase to 177% and 126% of controls, respectively) of *Hypericum* extracts. Another microdialysis study measuring 5-HT levels in the nucleus accumbens following administration of 1 g/kg p.o. of *Hypericum perforatum*–CO_2 extract reached the same conclusion [33]. This formulation contains a high (23.7%) concentration of hyperforin, which based on *in vitro* studies, was proposed to be responsible for the 5-HT reuptake inhibitory properties of the *Hypericum* extracts. Conventional SSRIs given systemically reduce the firing rate of the 5-HT neurons in the dorsal raphe nucleus as a consequence of the activation of the negative feedback loop via inhibitory somato-dendritic 5-HT_{1A} autoreceptors. However, no such effect was observed in awake cats following systemic administration of the *Hypericum* extract, Jarsin 300 given at 15–600 mg/kg p.o. [34], contrary to the effects induced by SSRIs fluoxetine and sertraline.

In summary, available microdialysis and electrophysiological findings indicate that the mode of action of *Hypericum* extract on 5-HT function *in vivo* differs from that of the conventional SSRIs.

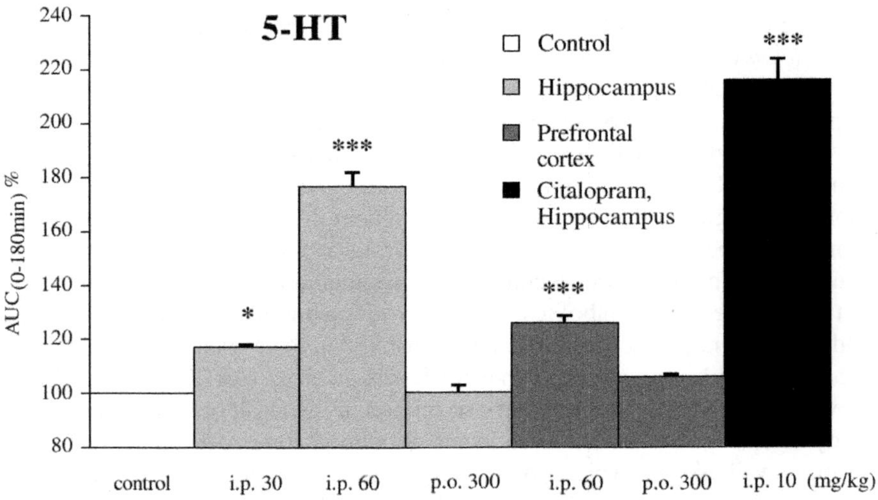

Figure 2. The effect of ethanolic extract of *Hypericum perforatum* on extracellular levels of serotonin in the prefrontal cortex and hippocampus of awake rat. *Hypericum* was given at doses of 30, 60 kg/kg i.p. and 300 mg/kg p.o. Citalopram (10 mg/kg i.p.) was chosen as a reference compound. (*) $P < 0.05$; (***) $P < 0.001$, ANOVA, followed by Fisher's PLSD-test. All other conditions as in Figure 1.

Dopamine

Stimulation of mesolimbic and mesocortical DA system is strongly implicated in reward and motivation-related behaviour, whereas DA hypofunction is associated with development of dysthymia, dysphoria and anhedonia, which often precede the symptoms of major depression. A large body of evidence suggests that antidepressant treatments enhance DA transmission and increase postsynaptic DA receptors sensitivity (for review, see [35]). In addition, chronic treatment with antidepressants modulates DA release, particularly in the PFC and increases the DA, D2 and D3 receptors sensitivity, whereas such treatment decreases D1 receptor number and sensitivity [35, 36].

Only limited information exists on the effects of *Hypericum* extracts or its constituents on DA function *in vivo* and *ex vivo*. The alcoholic extract of *Hypericum* was shown to be three times more potent to inhibit synaptosomal uptake of DA (IC_{50} 0.85 μg/ml) than 5-HT and more than five times more effective on DA uptake than that of NA [10]. Increased DA content was observed only in the rat diencephalon but not in the cortex following oral administration of the highest dose tested (500 mg/kg) of *Hypericum* extracts [13, 14] and chronic treatment for 2 and 8 weeks with *Hypericum* extract or hyperforin had no effect or even reduced the DA levels in the hypothalamus [16]. In microdialysis studies, the DA levels in the nucleus accumbens of *Hypericum*-treated rats were modestly increased compared with vehicle-treated rats [32, 33]. 3 weeks of treatment with *Hypericum perforatum* prevented the reduction of extracellular DA levels in the nucleus accumbens shell induced by chronic stress [37]. Hyperforin was shown to cause a minor increase in DA push–pull superfusates in the locus coeruleus [17].

A recent paper has examined the effects of *Hypericum* extract on extracellular DA levels in the limbic (striatum) and cortical (PFC) areas [29]. A summary of the results is shown in Figure 3. In the PFC, both the i.p and p.o. doses of *Hypericum* extract caused significant increases in DA levels, which were clearly higher than the corresponding values for NA and 5-HT (see Figs 1 and 2). The maximal increase of striatal DA levels was observed following a single 60 mg/kg i.p. dose, which also caused a significant motor stimulation in non-habituated rats [29]. These data support findings suggesting a role of DA in antidepressant action of *Hypericum* extracts based on altered levels of neuroendocrine responses known to be mediated by monoamines [38]. Thus, a single dose of methanolic extract of *Hypericum perforatum* to healthy volunteers caused a significant increase in plasma growth hormone and a decrease in prolactine levels, whereas cortisol values were unchanged [38]. A follow up study in rats [39] confirmed the initial observations in humans and proposed that the antidepressant effects of *Hypericum* extract (LI 160) may be mediated mainly via increased dopaminergic and serotonergic function.

In conclusion, the present *in vivo* data suggest that *Hypericum* extract have powerful stimulatory effects on DA transmission in cortico-limbic regions of the brain. However, it is possible that some effects may be related to certain active constituents of *Hypericum* such as flavonoids, which can exert, either alone or in synergy, stimulatory effects on mesolimbic and mesocortical dopamine pathways.

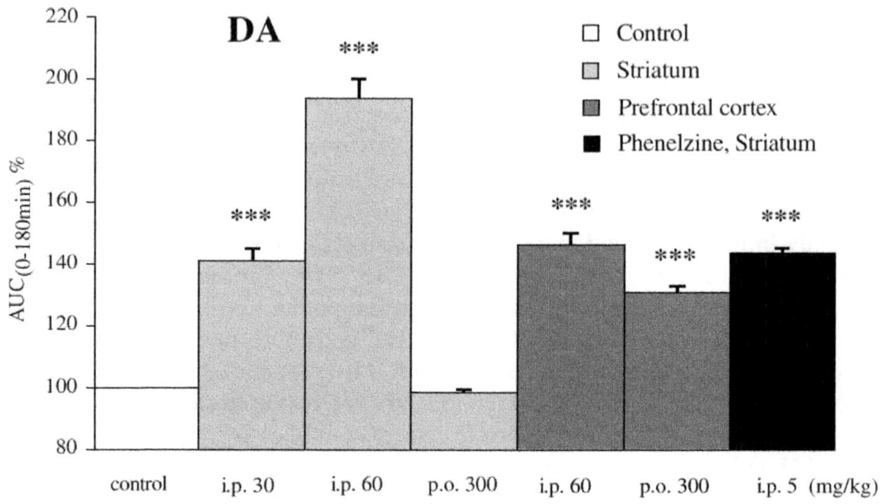

Figure 3. The effect of ethanolic extract of *Hypericum perforatum* on extracellular levels of dopamine in the prefrontal cortex and striatum of awake rat. *Hypericum* was given at doses of 30, 60 kg/kg i.p. and 300 mg/kg p.o. Phenelzine (5 mg/kg i.p.) was chosen as a reference compound. (***) $P < 0.001$, ANOVA, followed by Fisher's PLSD-test. All other conditions as in Figure 1.

Acetylcholine

Monitoring extracellular levels of acetylcholine (ACh) by microdialysis is frequently used as a neurochemical correlate to study the role of ACh in learning and memory. Behavioural studies in rats suggest that treatment with *Hypericum* extracts can improve spatial memory [40]. On the basis of *in vitro* data suggesting that inhibition of 5-HT reuptake by hyperforin is mediated via the influx of sodium ions [20], it was speculated also that other transmitter systems may be regulated by this mechanism. Thus, extracellular ACh levels were determined in the rat striatum following local infusion of hyperforin [41]. Hyperforin infusion at 10 µM or systemic administration (1–10 mg/kg i.p.) caused a significant increase in striatal ACh levels, whereas infusion of a higher dose (100 µM) caused a reduction of ACh release. Interestingly, this dose induced an increase in extracellular choline levels, indicating that hyperforin could interfere via effects on sodium channels, which in turn could modulate sodium-dependent high-affinity uptake mechanism of choline. The later electrophysiological study in mouse neuromuscular junction also indicates potentiation of ACh function by *Hypericum* extract, suggested to be mediated via inhibition of AChE activity [42].

It is concluded that further studies are necessary to evaluate the hypothesis that *Hypericum* extracts may improve cognitive function by increasing cholinergic neurotransmission in brain areas implicated in learning and memory, e.g., hippocampus and PFC.

Glutamate and GABA

Hyperforin, which is suggested to play a major role in the antidepressant effect of *Hypericum* extracts, was also shown to inhibit the uptake of tritiated glutamate and GABA in rat synaptosomal preparations [21, 43]. Interestingly, the lowest IC_{50} values measured for GABA (1.11 µg/ml) and Glu (2.87 µg/ml) in the extracts containing higher (>0.9%) levels of hyperforin were comparable or even lower than those required for inhibition of 5-HT and NA uptake [21]. These data were supported by a study on the effect of hyperforin on the release of amino acids from mouse cortical slices [44]. Thus, the presence of 5 µM hyperforin in the superfusion medium increased the efflux of aspartate, glutamate, serine, glycine and GABA but not taurine. In agreement with an earlier hypothesis on the mechanism of action of *Hypericum* and hyperforin [20], the authors proposed that hyperforin could increase the overflow of amino acids via facilitation of influx of sodium ions and mobilisation of intracellular calcium stores. There are no current data available, which support these *in vitro* findings in experiments carried-out *in vivo*. One study reports an increase, specifically of Glu (for more than 6 h!) but no other amino acid, sampled by push–pull perfusion in locus coeruleus of anaesthetised and acutely implanted rats following intraperitoneal injection of 10 mg/kg of hyperforin [17]. However, these data should be interpreted with caution in view of the strong experimental evidence that minimal invasive techniques such as acute implantation of the cannula or a microdialysis probe cause local trauma and result in abnormal levels of amino acid neurotransmitters [45]. Furthermore, an electrophysiological study in hippocampal slices of a guinea pig showed that the *Hypericum* extract but not hypericin and hyperforin exerted excitatory action on CA1 hippocampal neurons [46], suggesting that some other *Hypericum* constituents might be involved in this effect.

In summary, rather preliminary and inconsistent experimental evidence support the notion that *Hypericum* extract may increase glutamatergic and GABA-ergic neurotransmission via inhibition of uptake or stimulation of release of these neurotransmitters *in vivo*.

Conclusions

The *ex vivo* and *in vivo* neurochemical techniques were applied in order to examine the effects of St. John's Wort extracts on neurotransmitter release and metabolism, to elucidate mechanisms of action of *Hypericum* and its constituents and to compare these effects to those reported for current antidepressants such as SSRIs.

Based on available evidence summarized in this review we conclude the following:

- Results obtained under *in vitro* and *in vivo* conditions display striking differences.

- The mechanism of action of St. John's Wort extract and hyperforin is more complex than that ascribed to antidepressants SSRIs, NSRIs or MAOIs.
- In animal models, most synthetic antidepressants exert an immediate effect on tissue and extracellular monoamine content already following a single acute dose.
- The effects of *Hypericum* on inhibition of monoamine reuptake seem to be rather unspecific, probably mediated via changes in intracellular sequestration of sodium/calcium/proton ions. Such a mechanism may explain the observed affect on non-aminergic neurotransmitters such as ACh, Glu, GABA.
- Microdialysis data from our laboratory suggest that *Hypericum* extracts exert the most significant effects on the extracellular levels on monoamines in the order, DA > 5-HT > NA and in neuroanatomical areas relevant for the pathophysiology of depression (prefrontal cortex, hippocampus).
- The present findings indicate that further exploration of the mechanisms of action of St. John's Wort in depression should focus on DA and to some extent also 5-HT functions.

References

1 Lecrubier Y, Clerc G, Didi R, Kieser M (2002) Efficacy of St. John's Wort extract WS 5570 in major depression: a double-blind, placebo-controlled trial. *Am J Psychiatry* 159: 1361–1366

2 Barnes J, Anderson LA, Phillipson JD (2001) St John's Wort (*Hypericum perforatum* L.): a review of its chemistry, pharmacology and clinical properties. *J Pharm Pharmacol* 53: 583–600

3 Müller WE (2003) Current St John's Wort research from mode of action to clinical efficacy. *Pharmacol Res* 47: 101–109

4 Butterweck V (2003) Mechanism of action of St John's Wort in depression: what is known? *CNS Drugs* 17: 539–562

5 Coppen A (1967) The biochemistry of affective disorders. *Br J Psychiatry* 113: 1237–1264

6 Kehr J, Fuxe K, Ungerstedt U, Svensson TH (eds): (2003) *Monitoring Molecules in Neuroscience.* Karolinska University, Stockholm

7 Dawson LA, Nguyen HQ, Geiger A (1999) Effects of venlafaxine on extracellular concentrations of 5-HT and noradrenaline in the rat frontal cortex: augmentation via 5-HT1A receptor antagonism. *Neuropharmacology* 38: 1153–1163

8 Bymaster FP, Dreshfield-Ahmad LJ, Threlkeld PG, Shaw JL, Thompson L, Nelson DL, Hemrick-Luecke SK, Wong DT (2001) Comparative affinity of duloxetine and venlafaxine for serotonin and norepinephrine transporters *in vitro* and *in vivo*, human serotonin receptor subtypes, and other neuronal receptors. *Neuropsychopharmacology* 25: 871–880

9 Müller WE, Singer A, Wonnemann M, Hafner U, Rolli M, Schafer C (1998) Hyperforin represents the neurotransmitter reuptake inhibiting constituent of *Hypericum* extract. *Pharmacopsychiatry* 31 (Suppl 1): 16–21

10 Chatterjee SS, Bhattacharya SK, Wonnemann M, Singer A, Müller WE (1998) Hyperforin as a possible antidepressant component of *Hypericum* extracts. *Life Sci* 63: 499–510

11 Kientsch U, Burgi S, Ruedeberg C, Probst S, Honegger UE (2001) St. John's Wort extract Ze 117 (*Hypericum perforatum*) inhibits norepinephrine and serotonin uptake into rat brain slices and reduces 3-adrenoceptor numbers on cultured rat brain cells. *Pharmacopsychiatry* 34 (Suppl 1): S56–60

12 Butterweck V, Nahrstedt A, Evans J, Hufeisen S, Rauser L, Savage J, Popadak B, Ernsberger P, Roth BL (2002) *In vitro* receptor screening of pure constituents of St. John's Wort reveals novel interactions with a number of GPCRs. *Psychopharmacology* 162: 193–202

13 Calapai G, Crupi A, Firenzuoli F, Costantino G, Inferrera G, Campo GM, Caputi AP (1999)

Effects of *Hypericum perforatum* on levels of 5-hydroxytryptamine, noradrenaline and dopamine in the cortex, diencephalon and brainstem of the rat. *J Pharm Pharmacol* 51: 723–728
14 Calapai G, Crupi A, Firenzuoli F, Inferrera G, Squadrito F, Parisi A, De Sarro G, Caputi A (2001) Serotonin, norepinephrine and dopamine involvement in the antidepressant action of *Hypericum perforatum*. *Pharmacopsychiatry* 34: 45–49
15 Yu PH (2000) Effect of the *Hypericum perforatum* extract on serotonin turnover in the mouse brain. *Pharmacopsychiatry* 33: 60–65
16 Butterweck V, Bockers T, Korte B, Wittkowski W, Winterhoff H (2002) Long-term effects of St. John's Wort and hypericin on monoamine levels in rat hypothalamus and hippocampus. *Brain Res* 930: 21–29
17 Kaehler ST, Sinner C, Chatterjee SS, Philippu A (1999) Hyperforin enhances the extracellular concentrations of catecholamines, serotonin and glutamate in the rat locus coeruleus. *Neurosci Lett* 262: 199–202
18 Franklin M, Cowen PJ (2001) Researching the antidepressant actions of *Hypericum perforatum* (St. John's Wort) in animals and man. *Pharmacopsychiatry* 34 (Suppl 1): S29–37
19 Perovic S, Müller WE (1995) Pharmacological profile of *Hypericum* extract. Effect on serotonin uptake by postsynaptic receptors. *Arzneimittelforschung* 45: 1145–1148
20 Singer A, Wonnemann M, Müller WE (1999) Hyperforin, a major antidepressant constituent of St. John's Wort, inhibits serotonin uptake by elevating free intracellular Na^{+1}. *J Pharmacol Exp Ther* 290: 1363–1368
21 Wonnemann M, Singer A, Siebert B, Müller WE (2001) Evaluation of synaptosomal uptake inhibition of most relevant constituents of St. John's Wort. *Pharmacopsychiatry* 34 (Suppl 1): S148–151
22 Gobbi M, Valle FD, Ciapparelli C, Diomede L, Morazzoni P, Verotta L, Caccia S, Cervo L, Mennini T (1999) *Hypericum perforatum* L. extract does not inhibit 5-HT transporter in rat brain cortex. *Naunyn Schmiedebergs Arch Pharmacol* 360: 262–269
23 Roz N, Mazur Y, Hirshfeld A, Rehavi M (2002) Inhibition of vesicular uptake of monoamines by hyperforin. *Life Sci* 71: 2227–2237
24 Roz N, Rehavi M (2003) Hyperforin inhibits vesicular uptake of monoamines by dissipating pH gradient across synaptic vesicle membrane. *Life Sci* 73: 461–470
25 Bhattacharya SK, Chakrabarti A, Chatterjee SS (1998) Activity profiles of two hyperforin containing *Hypericum* extracts in behavioral models. *Pharmacopsychiatry* 31 (Suppl 1): 22–29
26 Cervo L, Rozio M, Ekalle-Soppo CB, Guiso G, Morazzoni P, Caccia S (2002) Role of hyperforin in the antidepressant-like activity of *Hypericum perforatum* extracts. *Psychopharmacology* 164: 423–428
27 Misane I, Ogren SO (2001) Effects of *Hypericum perforatum* (St. John's Wort) on passive avoidance in the rat: evaluation of potential neurochemical mechanisms underlying its antidepressant activity. *Pharmacopsychiatry* 34 (Suppl 1): S89–97
28 Butterweck V, Christoffel V, Nahrstedt A, Petereit F, Spengler B, Winterhoff H (2003) Step by step removal of hyperforin and hypericin: activity profile of different *Hypericum* preparations in behavioral models. *Life Sci* 73: 627–639
29 Yoshitake T, Iizuka R, Yoshitake S, Weikop P, Müller WE, Ogren SO, Kehr J (2004) *Hypericum perforatum* L (St. John's Wort) preferentially increases extracellular dopamine levels in the rat prefrontal cortex. *Br J Pharmacol* 142: 414–418
30 Yoshitake T, Kehr J, Yoshitake S, Fujino K, Nohta H, Yamaguchi M (2004) Determination of serotonin, noradrenaline, dopamine and their metabolites in rat brain extracts and microdialysis samples by column liquid chromatography with fluorescence detection following derivatization with benzylamine and 1,2-diphenylethylenediamine. *J Chromatogr B Analyt Technol Biomed Life Sci* 807: 177–183
31 Kehr J (1999) Monitoring chemistry of brain microenvironment: biosensors, microdialysis and related techniques. In: U Windhorst, H Johansson (eds): *Modern techniques in neuroscience research.* Springer-Verlag GmbH, Heidelberg. Chapter 41: 1149–1198
32 Rommelspacher H, Siemanowitz B, Mannel M (2001) Acute and chronic actions of a dry methanolic extract of *Hypericum perforatum* and a hyperforin-rich extract on dopaminergic and serotonergic neurones in rat nucleus accumbens. *Pharmacopsychiatry* 34 (Suppl 1): S119–126
33 Di Matteo V, Di Giovanni G, Di Mascio M, Esposito E (2000) Effect of acute administration of *Hypericum perforatum*-CO_2 extract on dopamine and serotonin release in the rat central nervous system. *Pharmacopsychiatry* 33: 14–18

34 Fornal CA, Metzler CW, Mirescu C, Stein SK, Jacobs BL (2001) Effects of standardized extracts of St. John's Wort on the single-unit activity of serotonergic dorsal Raphe neurons in awake cats: comparisons with fluoxetine and sertraline. *Neuropsychopharmacology* 25: 858–870
35 D'Aquila PS, Collu M, Gessa GL, Serra G (2000) The role of dopamine in the mechanism of action of antidepressant drugs. *Eur J Pharmacol* 405: 365–373
36 Willner P (1997) The mesolimbic dopamine system as a target for rapid antidepressant action. *Int Clin Psychopharmacol* 12 (Suppl 3): S7–14
37 Gambarana C, Tolu PL, Masi F, Rinaldi M, Giachetti D, Morazzoni P, De Montis MG (2001) A study of the antidepressant activity of *Hypericum perforatum* on animal models. *Pharmacopsychiatry* 34 (Suppl 1): S42–44
38 Franklin M, Chi J, McGavin C, Hockney R, Reed A, Campling G, Whale RW, Cowen PJ (1999) Neuroendocrine evidence for dopaminergic actions of Hypericum extract (LI 160) in healthy volunteers. *Biol Psychiatry* 46: 581–584
39 Franklin M, Chi JD, Mannel M, Cowen PJ (2000) Acute effects of LI 160 (extract of *Hypericum perforatum*, St John's Wort) and two of its constituents on neuroendocrine responses in the rat. *J Psychopharmacol* 14: 360–363
40 Widy-Tyszkiewicz E, Piechal A, Joniec I, Blecharz-Klin K (2002) Long term administration of *Hypericum perforatum* improves spatial learning and memory in the water maze. *Biol Pharm Bull* 10: 1289–1294
41 Buchholzer ML, Dvorak C, Chatterjee SS, Klein J (2002) Dual modulation of striatal acetylcholine release by hyperforin, a constituent of St. John's Wort. *J Pharmacol Exp Ther* 301: 714–719
42 Re L, Corneli C, Sturani E, Paolucci G, Rossini F, Leon OS, Martinez G, Bordicchia M, Tomassetti Q (2003) Effects of *Hypericum* extract on the acetylcholine release: a loose patch clamp approach. *Pharmacol Res* 48: 55–60
43 Wonnemann M, Singer A, Müller WE (2000) Inhibition of synaptosomal uptake of ^{3}H-Lglutamate and ^{3}H-GABA by hyperforin, a major constituent of St. John's Wort: the role of amiloride sensitive sodium conductive pathways. *Neuropsychopharmacology* 23: 188–197
44 Marsh WL, Davies JA (2002) The involvement of sodium and calcium ions in the release of amino acid neurotransmitters from mouse cortical slices elicited by hyperforin. *Life Sci* 71: 2645–2655
45 Clapp-Lilly KL, Roberts RC, Duffy LK, Irons KP, Hu Y, Drew KL (1999) An ultrastructural analysis of tissue surrounding a microdialysis probe. *J Neurosci Methods* 90: 129–142
46 Langosch JM, Zhou XY, Heinen M, Kupferschmid S, Chatterjee SS, Noldner M, Walden J (2002) St John's Wort (*Hypericum perforatum*) modulates evoked potentials in guinea pig hippocampal slices via AMPA and GABA receptors. *Eur Neuropsychopharmacol* 12: 209–216

St. John's Wort and its Active Principles in Depression and Anxiety
Edited by W. E. Müller

Efficacy in behavioural models of antidepressant activity

Carla Gambarana[1] and Daniela Giachetti[2]

[1] *Department of Neuroscience, Pharmacology Unit, University of Siena, Via Moro 4, 53100 Siena, Italy*
[2] *Department of Environmental Sciences, University of Siena, 53100 Siena, Italy*

Studies of the effects of *Hypericum perforatum* extracts in experimental models of depression, and of the possible mechanisms underlying these effects, greatly increased after evidence of efficacy of these extracts on mild and moderate forms of depression was offered by clinical studies and diffuse medical practice observations from German-speaking regions of Europe. During the past decade a large amount of data has demonstrated that *H. perforatum* extracts are active in diverse behavioural and biochemical models usually employed to screen antidepressant compounds [1]. Moreover, *in vivo* animal studies have suggested a role of the serotonergic, dopaminergic, noradrenergic, and opioidergic systems in the mechanisms of *H. perforatum* antidepressant activity [2, 3].

Studies were then aimed at identifying the possible active component(s) responsible for antidepressant activity. The majority of studies indicate that this antidepressant-like activity can be traced back to hyperforin (e.g., [4–8]). However, several studies provide evidence of hypericin activity in some depression models, either associated or not with flavonoids [3, 9, 10]. The issue is made more complex by the instability of some components during the preparation of the extract or when they are isolated from the extract, e.g., hyperforin [11], and by possible bioavailability problems when the oral administration of a purified compound, such as hypericin, pseudohypericin, or hyperforin, is used ([12] Gambarana et al., unpublished results). Thus, we do not yet have an unequivocal answer to the question of which of the *H. perforatum* extracts components is the main one responsible for antidepressant activity (at least in animal models). This information is crucial for determining the presence of the active principle in extracts and in the products derived from them that are commercially available worldwide, and to assess stability and shelf life of the active principle in the commercial products, under defined conditions. Only an accurate screening of the same product in multiple experimental models of depression could resolve this question.

Many behavioural models have been used to screen or study molecules that are potentially active in human depression. Some of these models show only a certain degree of *predictive validity*, as many of the compounds that are clinically efficacious in the treatment of depression are also active in these tests, but false positive and false negative results plague these models [13–15]. One example of these models that shows a behavioural effect of antidepressant drugs completely unrelated to an antidepressant activity is the reserpine test [16, 17]. These types of behavioural models are now largely dismissed, and researchers mostly utilise models that have *face* and *construct* as well as *predictive* validity [14]. The models of depression commonly used in rodents are based on animal exposure to an unavoidable stress, the more common being the Tail Suspension Test (TST) [18, 19], the Forced Swim Test (FST) (Fig. 1) [20], the Learned Helplessness model (LH) [21, 22] and derived models, e.g., the Escape Deficit (ED) (Fig. 2) [23, 24], and the Chronic Mild Stress (CMS) [25, 26]. In these models, the animal is faced with a stressful, inescapable (and unpredictable in the CMS) situation that has been interpreted as inducing a state of behavioural despair, akin to the hopelessness of depressive patients. The measured parameters are immobility (TST and FST), the lack of an avoidance response (LH and ED), the lack of a sucrose preference (CMS), or the lack of motivation to run in

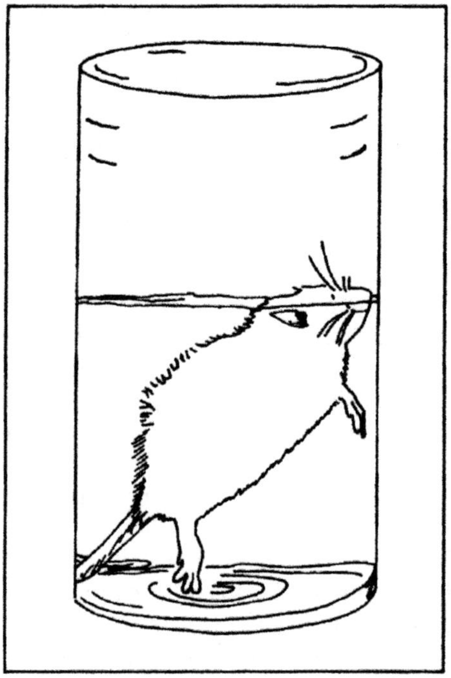

Figure 1. The forced swimming test (FST). In this test, rats or mice are placed in a cylinder filled with water at a temperature of 23–25 °C for 15 min. 24 h later animals are placed again in the cylinder for 5 min and their immobility is recorded [20]. Treatments are often done 24, 8–6, and 1 h before the 5 min test.

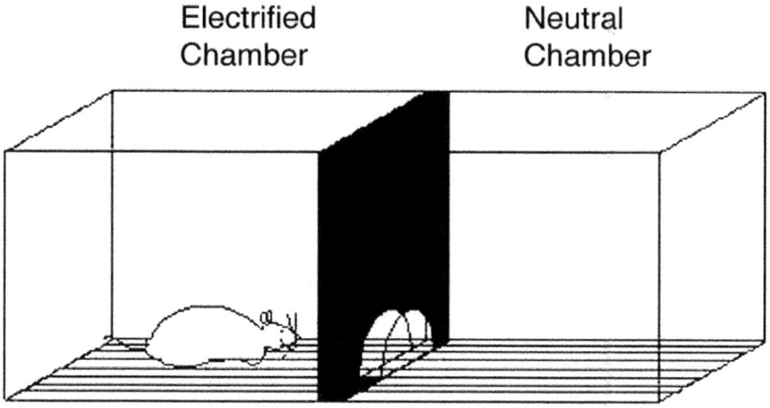

Figure 2. The escape deficit (ED). In the acute ED test, rats are exposed to an unavoidable stress (immobilisation and tail shocks) for 50 min. 24 h later they are tested in an escape test: they are placed in a shuttle-box and exposed to avoidable tail-shocks. The number of escapes is recorded [23]. Treatments are often done 14 days before unavoidable stress exposure. In the chronic ED after exposure to unavoidable stress and escape test, rats are exposed to a 3-week protocol of unavoidable stress exposure and they are then tested for escape [23]. Treatments begin after the first escape test and continue for 3 weeks to the last day of stress exposure.

a Y-maze for palatable food (stress-disruption of Vanilla-sugar sustained Appetitive Behaviour, VAB (Fig. 3) [27]). These models are all sensitive to clinically efficacious antidepressants, with some false negatives and false positives.

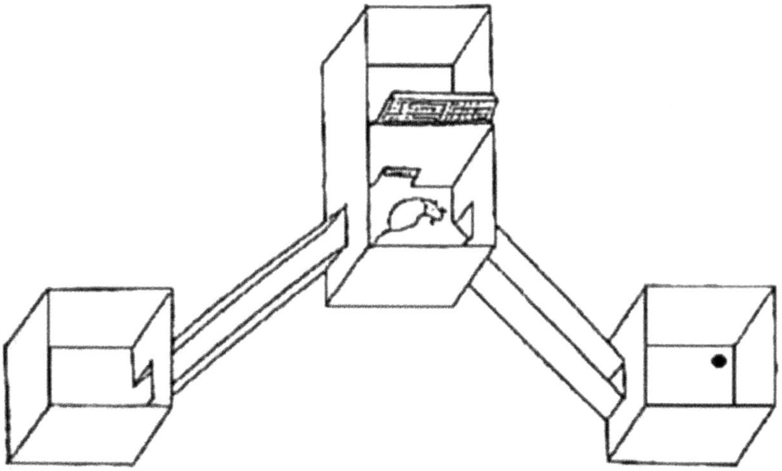

Figure 3. The stress-disrupted VAB. Satiated rats trained in a Y-maze to find a vanilla-sugar pellet easily learn in which arm the reward is located, and consistently run in the correct arm and consume the pellet. When rats are exposed to a chronic stress protocol and trained in the Y-maze, they never acquire the appetitive behaviour [27]. Treatments begin before the first stress exposure and continue as long as stress exposure and Y-maze training last.

Thus, only the activity, or lack of activity, in more than one model can be considered to be a sound indication of a possible antidepressant effect, or of the lack of a possible antidepressant effect. Moreover, it should be kept in mind that immobility in the TST and FST is reduced by an acute or subacute antidepressant treatment, while antidepressant clinical efficacy is only observed after long-term treatment. In the case of *H. perforatum* extracts, it is relevant to note that they have been found to be active in all of these behavioural models.

Efficacy of *Hypericum perforatum* extracts in behavioural models

H. perforatum extracts tested in models of depression are methanolic or ethanolic extracts, which in most cases originate from the European sources that prepare commercially available products or the raw material from which the commercial products are derived. It is interesting to note that a recent study that failed to confirm the activity of *H. perforatum* extracts in FST employed two extracts commercially available in Brazil [28]. One of the extracts was obtained from *H. perforatum* grown in Brazil and the other was imported from Europe. These data, even though obtained in a single model, the FST, again emphasize the issue of the need for *fingerprinting* the extract employed in basic or clinical studies, and for assessing the stability of its components in order to obtain meaningful and comparable results. However, we will be able to properly address this issue only when the active principle(s) have been identified. In most studies the extract was administered as a suspension by gavage.

The stress-induced models used to investigate *H. perforatum* effects can be divided into those that expose the animal to a single, acute unavoidable stress (TST, FST, LH, and acute ED), and those that employ repeated exposures to unavoidable stress (chronic ED, stress-disruption of VAB). Moreover, a further division could be made between stress-induced models that measure hyporeactivity, immobility time or number of failures to escape and models that measure a deficit in hedonia or in the motivation to obtain a natural reward (stress-disruption of VAB).

In the acute models, *H. perforatum* has been demonstrated to be active in reducing the immobility time after a single administration in mice in the TST [3], and in mice and rats after three administrations over 24 h in the FST [3, 29]. In the FST both a subacute treatment schedule (nine administrations over 4 days) and a 12-day repeated schedule of treatment have been shown to be effective in reducing immobility time in rats [10, 30]. Moreover, in the ED model the extract was able to prevent the escape deficit induced by stress exposure after either single or repeated administrations (a twice a day 14-day treatment) [2]. The range of doses used and shown to be active in acute models is quite large (from 50 mg/kg to 1 g/kg twice a day in repeated treatments in rats), and it varies with the model and the extract used.

Thus, either the acute, subacute, or repeated administrations of *H. perforatum* are efficacious in preventing the development of hyporeactivity after exposure to a single session of unavoidable stress. When a reduction in immobility is the measured parameter, a potentially confounding factor is the possible increase in locomotor activity induced by the drug used. Several studies have then looked for a possible motor stimulant activity of *H. perforatum* extracts and such an effect has not been found after either acute or repeated administration of a large range of doses (from 1.56 mg/kg i.p. to 250 mg/kg/die – 1 g/kg twice a day p.o.) [2, 3, 5]. On the contrary, a reduction in spontaneous locomotor activity has been observed in rats with doses of 6.25 mg/kg i.p. to 1 g/kg p.o., again after acute, subacute or repeated administration [2, 5]. A second potentially confounding factor when the stressor is a nociceptive stimulus is the possible analgesic activity of the studied compound. In order to rule out this possibility, the nociceptive threshold of *H. perforatum* treated rats has been determined with the tail-flick and the hot-plate tests. The nociceptive threshold is not modified in treated compared to control rats [2]. At variance with these data, the analgesic activity of *H. perforatum* was reported in mice with the tail-flick and tail-clip test [29, 31].

Additional information has been obtained in behavioural models by the use of receptor antagonists to interfere with the protective activity of the extract. A study showed the antagonism of the protective effect of repeated *H. perforatum* treatment on the acute ED model in rats after the administration of a selective 5-HT_{1A} receptor antagonist (WAY 100635), or a DA D_1 receptor antagonist (SCH 23390) [2]. In other studies that employed the FST in rats, antagonism of the effect of *H. perforatum* subacute treatment was observed after the administration of a DA D_2 receptor antagonist (sulpiride or haloperidol) [3]. In mice, an inhibition of *H. perforatum* anti-immobility effect in the TST was reported after administration of an opiate receptor antagonist, naloxone, or when α-methyl-para-tyrosine was injected 3 h before the extract [3]. Thus, a dopaminergic component seems to be present in the antidepressant effect of the extract in these different animal models, but other receptor systems are also involved, probably with distinct relevance in different species and/or behavioural tests.

In the models induced by chronic stress exposure, *H. perforatum* (1 g/kg twice a day p.o.) was able to restore a normal escape response in chronic ED after twice a day administration for 21 days, an effect shared with clinically efficacious antidepressants [2]. In the VAB model, where rats exposed to chronic stress do not learn an instrumental appetitive behaviour, the effect of long-term *H. perforatum* administration was compared to the effect of fluoxetine and imipramine, established antidepressant drugs [2]. *H. perforatum* and fluoxetine were administered to control rats and to rats exposed to chronic stress after the induction of an escape deficit. In this experiment, *H. perforatum* and fluoxetine were used at doses of 1 g/kg twice a day p.o. and 5 mg/kg/day i.p., respectively. Figure 4 shows that neither of the drugs impaired the ability of rats to learn the appetitive behaviour, and they allowed stressed rats to learn and perform the test as efficiently as control rats.

Most of the studies of models of depression induced by acute or chronic stress exposure used one or more antidepressant drugs of established efficacy as

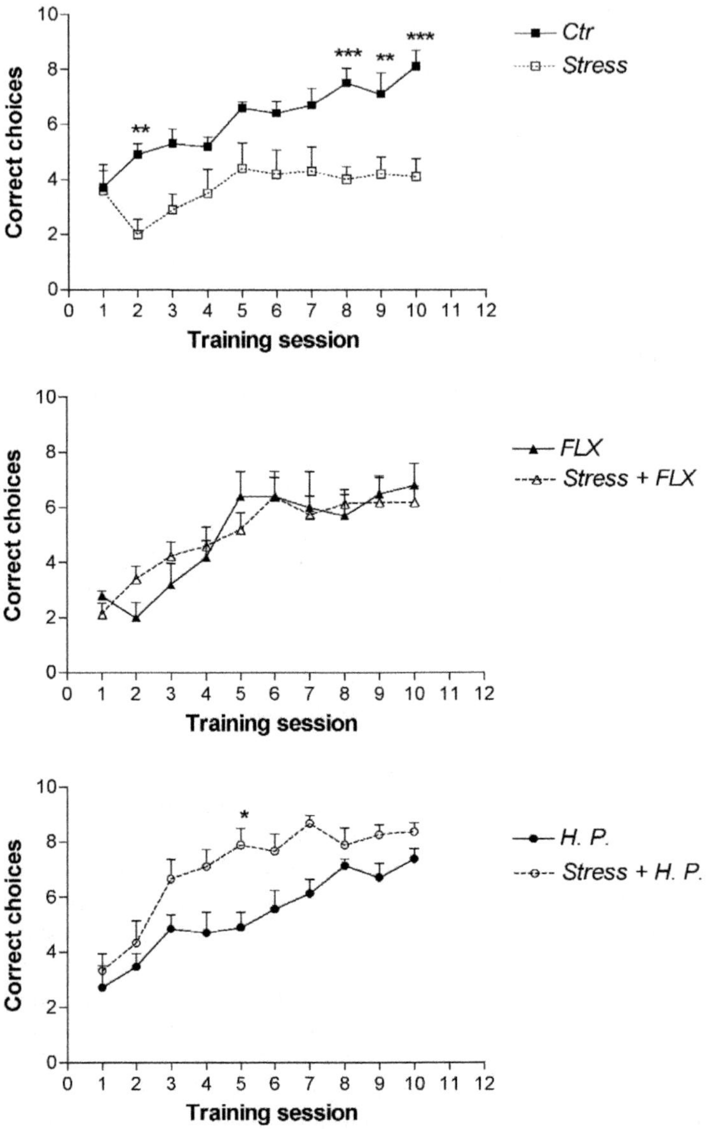

Figure 4. VAB performance of saline (Ctr), fluoxetine (FLX), and *H. perforatum* (H.P.) treated rats, exposed or not exposed to chronic stress during Y-maze training. Rats received daily saline (1 ml/kg, twice a day), fluoxetine (5 mg/kg/day, i.p.), or *H. perforatum* (1 mg/kg twice a day, p.o.) and were exposed to unavoidable stress and Y-maze training on alternate days. The number of correct choices, i.e., the number of times that a rat entered the maze arm containing the reward (a vanilla sugar pellet) were recorded. Data are expressed as mean ± S.E.M. At the 10th training session, the number of correct choices of saline-treated rats exposed to stress (*Stress*) was significantly lower than the number of correct choices of the groups not exposed to stress (*Ctr, FLX, H. P.*) and of the groups exposed to stress while receiving fluoxetine or *H. perforatum* (*Stress + FLX, Stress + H. P.*) ($p < 0.001$, two-way ANOVA followed by Bonferroni's test). *** $p < 0.001$, ** $p < 0.01$, * $p < 0.05$ compared to the respective stressed group performance.

a positive control for the activity of *H. perforatum* extract in that model. Thus, the conclusion is that in the most commonly used animal models, *H. perforatum* extracts show an activity comparable to that of classical antidepressants.

Efficacy of components of *Hypericum perforatum* extracts in behavioural models

Many studies have attempted to identify one component, or a class or subclass of components, of the extracts as the active principle that has activity in depression models. Two complementary strategies have been used that could yield partially different information. The activity of extracts with different contents of one or more compounds was compared, and thus the increased or decreased efficacy compared to the standard extract was attributed to the increased or reduced content of one or more components in the extract. The second strategy was to compare the efficacy of the extract with the activity of a single purified component. While the second strategy could be considered as proof of the antidepressant activity of a certain *H. perforatum* component, sometimes the results are conflicting, as activity or lack of activity can be observed depending on the route of administration of the purified component, or on the behavioural test used. These data suggest that when a single component, which is assumed to be the one most likely responsible for antidepressant activity, is purified from the extract, it may need to be slightly modified in order to maintain the stability and oral bioavailability that it has in the extract. In the future, data regarding derivatives of natural compounds will be available and they may clarify the issue of the active antidepressant principle present in *H. perforatum* extracts.

Hyperforin and hyperforin derivatives

Several studies demonstrated that the antidepressant efficacy of *H. perforatum* extracts in different behavioural tests correlates with their hyperforin content. Chatterjee et al. showed the efficacy in the FST and LH of repeated oral administration of an ethanolic extract (hyperforin content 4.5%) or a CO_2 extract (hyperforin content 38.8%) in rats [6]. The dose equieffective to 10 mg/kg/day i.p. of imipramine was 30 mg/kg/day for the CO_2 extract and 300 mg/kg/day for the ethanolic extract. Moreover, in both tests efficacy was correlated to hyperforin content, but not to the presence or proportion of other constituents [6]. Similar results were reported in another study where the FST and other models predictive of antidepressant activity (e.g., reserpine-induced syndrome) were used [4]. Two extracts with different hyperforin contents (0.5% and 4.5%) were used by Cervo et al. in the FST. After three i.p. administrations over 24 h only the extract containing 4.5% hyperforin reduced immobility in rats, and this effect was accompanied by a decrease in locomotor activity in the open field test [5]. A CO_2 *H. perforatum* extract, enriched five times in hyperforin compared to the standard extract, was tested in the acute ED model [7]. The CO_2 extract was administered to rats acutely p.o. before stress exposure at doses ranging from 50–200 mg/kg. The dose-response curve showed that this extract was five times more potent compared to the standard extract. Moreover, it retained its protective effect after a 14-day administration at the dose of 100 mg/kg twice a day, i.e., tolerance to the antidepressant-like effect of the CO_2 extract did not develop [7].

A slightly different approach for verification of the hypothesis that hyperforin is probably responsible for *H. perforatum* antidepressant activity was chosen by Usai and co-workers, who used an extract of a subspecies of *H. perforatum, H. perforatum angustifolium* [8]. In the *H. perforatum angustifolium* extract used, the concentrations of hyperforin, I3,II8 biapigenin, and quercetin were two-fold that those present in a standard extract of *H. perforatum*, and rutin was almost absent [8]. The extract of *H. perforatum angustifolium* showed a protective activity on the development of acute ED after acute or repeated (15 day) oral administration, and it restored normal reactivity in rats

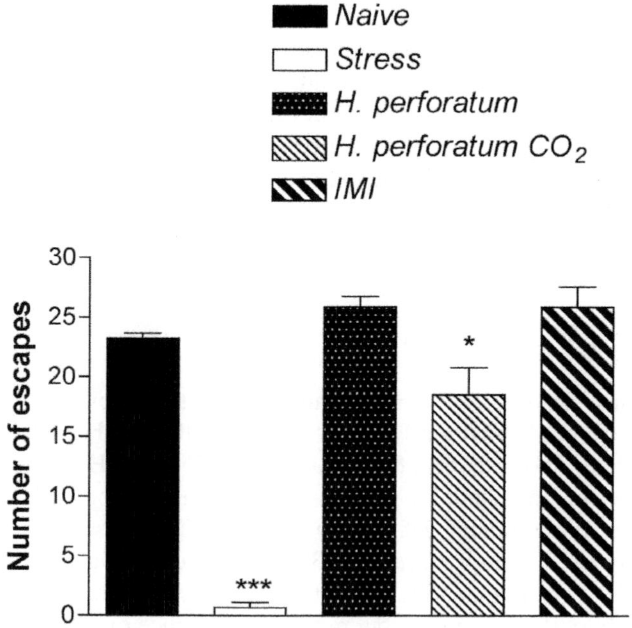

Figure 5. Protective effect of repeated treatments on escape deficit development. Rats received twice a day for 2 weeks: saline, 1 ml/kg (*Stress*), *H. perforatum* standard extract, 1 g/kg p.o. (*H. perforatum*), *H. perforatum* CO_2 extract, 100 mg/kg i.p. (*H. perforatum* CO_2), or imipramine, 5 mg/kg i.p. (*IMI*). 18 h after the last drug treatment, rats were exposed to unavoidable stress and 24 h later to the escape test. A group of saline-treated rats was only exposed to the escape test (*Naive*). Data are expressed as mean ± S.E.M. *** $p < 0.001$ compared to the *Naive, H. perforatum, H. perforatum* CO_2, and *IMI* groups; * $p < 0.05$ compared to the *H. perforatum* and *IMI* groups (one-way ANOVA followed by Bonferroni's test).

exposed to chronic stress (chronic ED). A significant protective activity of this extract was already present at a dose eight times lower than that necessary to produce a similar activity in the same tests when using *H. perforatum* extract [8]. Since the comparison of the composition of the two extracts used showed that *H. perforatum angustifolium* had a concentration of hyperforin double that of *H. perforatum*, while the hypericin content was similar, these data support the role of hyperforin in the antidepressant-like activity of *H. perforatum* and they suggest that hypericin does not contribute to the higher biological activity of *H. perforatum angustifolium*. Moreover, in these experimental models the flavonoid rutin does not seem to be necessary for *H. perforatum* activity, while its presence in at least a threshold concentration seems to be essential for the antidepressant activity of *H. perforatum* in the FST [12].

Thus, these data indicate that *H. perforatum* extracts enriched in hyperforin show an efficacy in experimental models of depression that is correlated to their hyperforin content, and this conclusion is strengthened by the results of studies that examined the effect of purified hyperforin. In the FST, after subacute oral administration at a dose of 20 mg/kg/die, hyperforin shows an efficacy comparable to that of imipramine [32]. Acute i.p. administration of hyperforin trimethoxybenzoate (IDN 5491) protects rats in the acute ED model at a dose of 25 mg/kg [7] and it maintains its protective activity on ED development after a 3-week repeated treatment. The active doses were lower than those that were efficacious when a hyperforin-enriched extract was used in the same tests [33]. However, we have also observed the lack of a protective effect of hyperforin trimethoxybenzoate when the same compound was administered orally (Tab. 1), thus suggesting that some component present in the

Table 1. Effect of acute hyperforin trimethoxybenzoate (IDN 5491) administration on stress-induced escape deficit[1]

Group	*n*	Number of escapes
Naive	12	25.3 ± 0.9
Ctr	12	1.2 ± 0.5***
IDN 5491 6.25 mg/kg i.p.	8	19.1 ± 3.2
IDN 5491 12.5 mg/kg i.p.	8	16.1 ± 0.9
IDN 5491 25 mg/kg i.p.	8	27.6 ± 1.1**
IDN 5491 12.5 mg/kg p.o.	6	1.4 ± 0.8***
IDN 5491 25 mg/kg p.o.	6	4.2 ± 1.3***
IDN 5491 100 mg/kg p.o.	6	5.2 ± 1.4***
IDN 5491 200 mg/kg p.o.	6	4.7 ± 1.6***

[1] Rats received hyperforin trimethoxybenzoate (IDN 5491) 30 min (i.p. administration) or 60 min (p.o. administration) before exposure to unavoidable stress; the control groups received saline (1 lm/kg i.p. or p.o., *Ctr*). A group of saline-treated rats was only exposed to the escape test (*Naive*). Data are expressed as mean ± S.E.M. *** $p < 0.001$ compared to the *Naive* group and the *IDN 5491* i.p. groups; ** $p < 0.01$ compared to the *IDN 5491* 12.5 mg/kg i.p. group (one-way Anova followed by Bonferroni's test).

extract may increase hyperforin bioavailability (Gambarana et al., unpublished results). A stable salt of the phloroglucinol derivative, hyperforin dicyclohexylammonium, administered i.p. three times over 24 h reduces immobility in the FST in rats, without modifying locomotor activity in the open field test [5]. Moreover, hyperforin dicyclohexylammonium exerts its protective effect on the consequences of unavoidable stress exposure through a central mechanism, as it was highly effective in the acute ED model after a 14-day intracerebroventricular infusion at a dose of 12 µg/kg/die (Fig. 6) (Gambarana et al., unpublished results). On the other hand, in the same animal model a de-hyperforinated extract was inactive after acute or repeated administration (Gambarana et al., unpublished results).

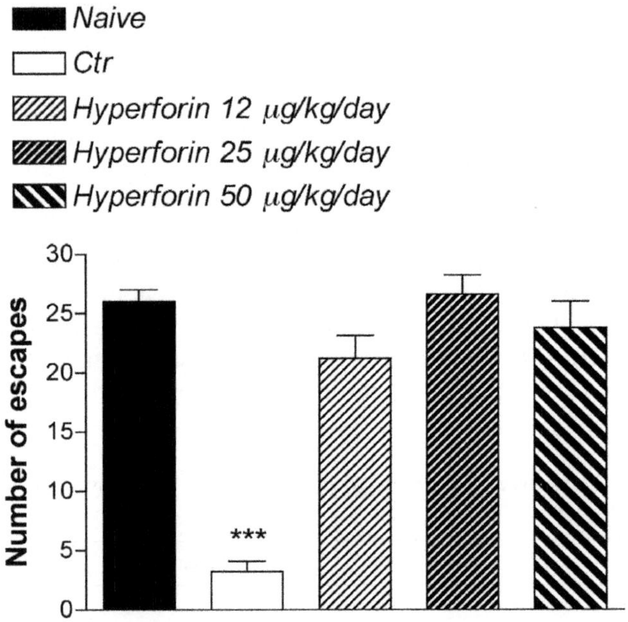

Figure 6. The protective effect of hyperforin on escape deficit development is centrally mediated. Rats were implanted subcutaneously with osmotic minipumps (Alzet®) under ether anaesthesia and were infused for 2 weeks with: saline (*Naive* and *Ctr* groups), or hyperforin cyclohexylammonium at the doses of 12, 25 or 50 µg/kg/day (*Hyperforin*). 18 h after the end of infusion, rats were exposed to unavoidable stress and 24 h later to the escape test. A group of saline-infused rats was only exposed to the escape test (*Naive*). Data are expressed as mean ± S.E.M. *** $p < 0.001$ compared to the *Naive* and *Hyperforin* groups (one-way ANOVA followed by Bonferroni's test).

Conclusions

The data in this Chapter are a summary of a significant part of the abundant literature on the subject and as a whole they confirm the efficacy of *H. perforatum* extracts in animal models of depression. Moreover, substantial evidence supports the hypothesis that hyperforin is probably the active principle with antidepressant activity in *H. perforatum* extract, although there are results indicating that other components, like hypericin, may be active in some experimental model (FST). It is worth noting that some studies reported that other components in the flavonoid class are necessary, or helpful, for observing hypericin efficacy in the FST. These results have been interpreted as related to the ability of flavonoids to improve the solubility or the passage through biological membranes of hypericin. Our observation that hyperforin is active when administered i.p., but that it fails to protect rats from the sequelae of stress exposure when administered orally, suggest that a similar phenomenon could also be true for the phloroglucinol derivative, which may require the presence of other components, or chemical modifications, in order to be adequately stable and to maintain good bioavailability.

We would like to remind the reader that other behavioural tests that we did not take into consideration may detect some other activities, such as an anxiolytic effect, that may strengthen the potential antidepressant effects of a compound. The potential anti-anxiety effect of *H. perforatum* extract or some extract components has been examined in animal models, although less extensively than the antidepressant-like effect. A different chapter of this volume is in fact dedicated to this issue.

The administration of *H. perforatum* extracts has also been demonstrated to be efficacious in a completely different model, which may be considered indirectly related to the depression or anxiety models, i.e., genetically selected alcohol preferring rat strains. In fact, in different strains of ethanol preferring rats, *H. perforatum* extract decreases ethanol consumption after acute [34, 35] or repeated administration [30, 35].

Thus, the challenge remains of identifying the main component responsible for *H. perforatum* extract activity in behavioural models of depression and to define the requirements for the preservation of such activity. We believe that only when basic research has demonstrated that this has been fulfilled, we will be able to properly design clinical studies aimed at assessing the efficacy and safety of *H. perforatum* extract in the treatment of depression.

Acknowledgements
The authors wish to thank Ms. Colleen Pisaneschi for language editing of the manuscript and INDENA S.p.A. for supplying the different *H. perforatum* extracts and hyperforin derivatives.

References

1 Müller WE (2003) Current St. John's Wort research from mode of action to clinical efficacy. *Pharmacol Res* 47: 101–109
2 Gambarana C, Ghiglieri O, Tolu P, De Montis MG, Giachetti D, Bombardelli E, Tagliamonte A (1999) Efficacy of an *Hypericum perforatum* (St. John's Wort) extract in preventing and reverting a condition of escape deficit in rats. *Neuropsychopharmacol* 21: 247–257
3 Butterweck V, Wall A, Liefländer-Wulf U, Winterhoff H, Nahrstedt A (1997) Effects of the total extract and fractions of *Hypericum perforatum* in animal assays for antidepressant activity.

Pharmacopsychiatry 30 (Suppl): 117–124
4 Bhattacharya S-K, Chakrabarti A, Chatterjee SS (1998) Activity profiles of two hyperforin-containing *Hypericum* extracts in behavioral models. *Pharmacopsychiatry* 31 (Suppl 2): 22–29
5 Cervo L, Rozio M, Ekalle-Soppo CB, Guiso G, Morazzoni P, Caccia S (2002) Role of hyperforin in the antidepressant-like activity of *Hypericum perforatum* extracts. *Psychopharmacology* 164: 423–428
6 Chatterjee SS, Bhattacharya SK, Wonnemann M, Singer A, Müller WE (1998) Hyperforin as a possible antidepressant component of Hypericum extracts. *Life Sci* 63: 499–510
7 Gambarana C, Tolu P, Masi F, Rinaldi M, Giachetti D, Morazzoni P, De Montis MG (2001) A study of the antidepressant activity of *Hypericum perforatum* on animal models. *Pharmacopsychiatry* 34 (Suppl): 42–44
8 Usai M, Leggio B, Grappi S, Nanni G, Gambarana C, Tolu P, Giachetti D, De Montis MG (2003) *Hypericum perforatum* subspecies *angustifolium* shows a protective activity on the consequences of unavoidable stress exposure at lower doses than *Hypericum perforatum perforatum*. *Pharmacopsychiatry* 36: 283–287
9 Butterweck V, Petereit F, Whinterhoff H, Nahrstedt A (1998) Solubilized hypericin and pseudohypericin from *Hypericum perforatum* exert antidepressant activity in the forced swimming test. *Planta Med* 64: 291–294
10 Butterweck V, Korte B, Winterhoff H (2001) Pharmacological and endocrine effects of *Hypericum perforatum* and hypericin after repeated treatment. *Pharmacopsychiatry* 34 (Suppl): 2–7
11 Erdelmeier CAJ (1998) Hyperforin, possibly the major non-nitrogenous secondary metabolite of *Hypericum perforatum* L. *Pharmacopsychiatry* 31 (Suppl): 2–6
12 Nöldner M, Schötz K (2002) Rutin is essential for the antidepressant activity of *Hypericum perforatum* extracts in the forced swimming test. *Planta Med* 68: 577–580
13 Willner P (1990) Animal models of depression: an overview. *Pharmacol Ther* 45: 425–455
14 Willner P (1991) Animal models as simulations of depression. *Trends Pharmacol Sci* 12: 131–136
15 Willner P (1995) Animal models of depression: validity and applications. *Adv Biochem Psychopharmacol* 49: 19–41
16 Dhawan KN, Jaju BP, Gupta GP (1970) Validity of antagonism of different effects of reserpine as test for anti-depressant activity. *Psychopharmacologia* 18: 94–98
17 Bourin M (1990) Is it possible to predict the activity of a new antidepressant in animals with simple psychopharmacological tests? *Fundam Clin Pharmacol* 4: 49–64
18 Steru L, Chermat R, Thierry B, Simon P (1985) The tail suspension test: a new method for screening antidepressants in mice. *Psychopharmacology* 85: 367–370
19 Steru L, Chermat R, Thierry B, Mico JA, Lenegre A, Steru M, Simon P, Porsolt RD (1987) The automated tail suspension test: a computerized device which differentiates psychotropic drugs. *Prog Neuropsychopharmacol Biol Psychiatry* 11: 659–671
20 Porsolt RD, LePichon R, Jalfre R (1977) Depression: a new animal model sensitive to antidepressant treatments. *Nature* 266: 730–732
21 Overmier JB, Seligman MEP (1967) Effects of inescapable shock upon subsequent escape and avoidance learning. *J Comp Physiol Psychol* 63: 23–33
22 Sherman AD, Sacquitine JL, Petty F (1982) Specificity of the learned helplessness model of depression. *Pharmacol Biochem Behav* 16: 449–454
23 Gambarana C, Scheggi S, Tagliamonte A, Tolu P, De Montis MG (2001) Animal models for the study of antidepressant activity. *Brain Res Prot* 7: 11–20
24 Meloni D, Gambarana C, De Montis MG, Dal Prà P, Taddei I, Tagliamonte A (1993) Dizocilpine antagonizes the effect of chronic imipramine on learned helplessness in rats, *Pharmacol Biochem Behav* 46: 423–426
25 Papp M, Moryl E, Willner P (1996) Pharmacological validation of the chronic mild stress model of depression. *Eur J Pharmacol* 296: 129–136
26 Papp M, Willner P, Muscat R (1991) An animal model of anhedonia: attenuation of sucrose consumption and place preference conditioning by chronic unpredictable mild stress. *Psychopharmacology* 104: 255–259
27 Ghiglieri O, Gambarana C, Scheggi S, Tagliamonte A, Willner P, De Montis MG (1997) Palatable food induces an appetitive behaviour in satiated rats which can be inhibited by chronic stress. *Behav Pharmacol* 8: 619–628
28 Guilhermano LG, Ortiz L, Ferigolo M, Barros HMT (2004) Commercially available *Hypericum perforatum* extracts do not decrease immobility of rats in the forced swimming test. *Prog*

Neuropsychopharmacol Biol Psychiatry 28: 49–55
29 Öztürk Y (1997) Testing the antidepressant effects of *Hypericum* species on animal models. *Pharmacopsychiatry* 30 (Suppl): 125–128
30 De Vry J, Maurel S, Schreiber R, de Beun R, Jentzsch KR (1999) Comparison of *Hypericum* extracts with imipramine and fluoxetine in animal models of depression and alcoholism. *Eur Neuropsychopharmacol* 9: 461–468
31 Öztürk Y, Aydin S, Beis R, Baser KHC, Berberoglu H (1995) Effects of *Hypericum perforatum* L. and *Hypericum calycinium* L. extracts on the central nervous system in mice. *Pythomedicine* 3: 137–144
32 Chatterjee SS, Nöldner M, Koch E, Erdelmeier C (1998) Antidepressant activity of *Hypericum perforatum* and hyperforin: the neglected possibility. *Pharmacopsychiatry* (Suppl) 31: 7–15
33 Gambarana C, Grappi S, Nanni G (2001) Dose-dependent antidepressant activity of hyperforin in rats. *Soc Neurosci Abstr* 27: 975.12
34 Rezvani AH, Overstreet DH, Yang Y, Clark E Jr, (1999) Attenuation of alcohol intake by extract of *Hypericum perforatum* (St. John's Wort) in two different strains of alcohol-preferring rats. *Alcohol Alcohol* 34: 699–705
35 Panocka I, Perfumi M, Angeletti S, Ciccocioppo R, Massi M (2000) Effects of *Hypericum perforatum* extract on ethanol intake, and on behavioral despair: A search for the neurochemical systems involved. *Pharmacol Biochem Behav* 66: 105–111

Edited by W. E. Müller

Comparative preclinical antidepressant activity of isolated constituents

Michael Nöldner

Dr. Willmar Schwabe GmbH & Co KG, Willmar-Schwabe-Straße 4, 76227 Karlsruhe, Germany

Introduction

Hydroalcoholic extracts of *Hypericum perforatum* L. (St. John's Wort) are effective antidepressants in patients, and also show antidepressant-like action in laboratory animals. Intensive phytochemical and pharmacological investigations have enabled the identification of a series of bioactive compounds which are constituents of the crude *Hypericum* extract [1]. The phloroglucinol derivative hyperforin has become a subject of particular interest, as it is a potent inhibitor of neuronal monoamine reuptake [2, 3]. However, in contrast to most synthetic antidepressants, hyperforin inhibits not only the uptake of serotonin, noradrenaline and dopamine, but also that of GABA and L-glutamate [4]. There is some evidence that the molecular mechanism involved in this broad-spectrum effect is based on an elevation of the intracellular Na^+ concentration [4, 5]. Different flavonol derivatives and biflavones have potent antidepressant activities in the "Forced Swimming Test" (FST), an animal model of depression [6, 7]. Interestingly, some of these compounds show activity over only a small dose range. Below or above these dosages, they become inactive. Besides this direct antidepressant-like activity of the flavones, a recent publication suggested that *Hypericum* extracts with no, or only very low, content of the flavonol rutin are completely inactive in the FST, while extracts containing rutin in sufficient amounts are always active [8]. Finally, the naphthodianthrone hypericin, the compound in *Hypericum* responsible for the phototoxic activity of the plant, shows antidepressant-like effects in animals, and may be involved in the regulation of the hypothalamic–pituitary–adrenal (HPA) axis, which is disturbed in depressed patients [9, 10]. It therefore appears clear that various constituents of *Hypericum perforatum* contribute in different ways to the antidepressant activity of the total extract. Although so much research has been done to identify and characterise the active constituents in the extract, there are still many questions about their efficacy and modes of action.

Phloroglucinols

The phloroglucinol content can exceed up to 5% in the fresh herb. Although the acylphloroglucinol compound hyperforin is the major non-nitrogenous secondary metabolite in *Hypericum perforatum*, for a long time it was not considered to be a pharmacologically relevant constituent, except in terms of its antimicrobial properties, which were described about 30 years ago [11]. Recently, however, these antibiotic-like effects were re-evaluated, and the influence of hyperforin and/or *Hypericum* extracts on resistance development in *Staphylococcus aureus* was investigated [12, 13]. In recent years, the preclinical efficacy of hyperforin has been extensively investigated. Numerous different pharmacological and biochemical effects were found in various *in vitro* and *in vivo* models. Some of the data generated in these studies clearly indicate that hyperforin not only contributes to the antidepressant activity of *Hypericum perforatum* extract but also shows anxiolytic and antidementia activities (see Chapter by S. Chatterjee in this book).

In vitro effects of hyperforin

Isolated organs

The first pharmacological activities recorded for hyperforin, apart from antimicrobial effects, were detected in peripheral organs using different *in vitro* models. An ethanolic *Hypericum* extract containing 4.5% hyperforin, a CO_2-extract containing 38.8% hyperforin, and pure hyperforin were shown to strongly inhibit contractions induced by serotonin, histamine, bradykinine, nicotine, substance-P and acetylcholine in guinea pig ileum preparations [14]. It was shown that the inhibition of serotonin-induced contractions correlated with hyperforin content, indicating that this phloroglucinol has serotonin-modulating properties. In other experiments conducted on neuronal membranes isolated from guinea pigs, hyperforin influenced membrane fluidity in a way that probably contributes to its pharmacological properties [15].

Uptake and release

Pure hyperforin and hyperforin-enriched extracts have been found to inhibit the reuptake of serotonin, noradrenaline, dopamine, GABA and L-glutamate in mice brain synaptosomes [2, 16], and of serotonin in human blood platelets [17].

Electrophysiology

In electrophysiological investigations using the patch-clamp technique in the whole cell configuration, it was shown that hyperforin at nanomolar concentrations induces significant inhibition of various ion channels known to be involved in neurotransmitter reuptake and release mechanisms. As a result of this study, it was shown that the P-type calcium channels in membranes of rat cerebellar Purkinje neurones are modulated by hyperforin via interaction with calmodulin or through a calmodulin-activated pathway involving secondary messengers [18–20]. Hyperforin has inhibitory effects on ligand-operated AMPA-, NMDA-, and GABA-gated currents, as well as voltage-dependent Na^+, K^+ and Ca^{2+} channels [20, 21]. The prevention of Ca^{2+} overloading by means of ligand- and voltage-dependent Ca^{2+}- permeable channel inhibition by hyperforin could be an important factor in the therapeutic efficacy of hyperforin [19].

Miscellaneous

In receptor binding studies, it was shown that hyperforin inhibits binding to the μ-, κ and δ-opioid as well as to 5-HT_6 and 5-HT_7 receptors to a relatively large extent in the micromolar range [22]. In a recently published study, it was shown that hyperforin has a considerable affinity for dopamine D_1 and D_5 receptors and for the noradrenaline transporter [23]. A hyperforin enriched CO_2 extract and a methanolic extract with lower hyperforin content showed only weak MAO-A or MAO-B inhibiting properties, indicating that hyperforin is not responsible for these effects [2].

In vivo effects of hyperforin

Besides its *in vitro* activities, hyperforin shows anti-serotonergic and antidepressant-like effects in different *in vivo* models. Serotonergic *in vivo* effects were investigated in rats using the method described by Saxena and Lawang [24]. In this test model, the heart rate is continuously monitored and a transient bradycardia is induced by intravenous infusion of serotonin (Bezold-Jarisch Reflex). This bradycardia, which is specifically mediated through vagus nerve 5-HT_3 receptors, is inhibited by oral administration of 10 mg/kg pure hyperforin [14]. It is suggested that antagonism of 5-HT_3-receptors produces antidepressant and anxiolytic-like effects [25], and that such effects could contribute to the antidepressant activity of *Hypericum* extracts [14]. The potential antidepressant activity of hyperforin or hyperforin-enriched extracts was investigated in a number of animal models using mice and rats. One of the most accepted animal models for detecting such activities is the FST developed by Porsolt et al. [26]. There is a significant correlation between the clinical activity of an antidepressant and its ability to reduce the immobility time in the FST [27].

After oral administration of 20 mg/kg hyperforin, a clear reduction in immobility time, comparable to the effect of the tricyclic antidepressant imipramine, was found in rats [14, 28]. In a recent publication, a stable hyperforin salt (dicyclohexylammonium) and two *Hypericum* extracts with different hyperforin content (0.5% and 4.5%) were investigated in the FST in rats. The extract with the higher hyperforin content and the hyperforin salt alone produced a marked reduction in the time spent immobile after intraperitoneal application of very low dosages. These effects correlated with the plasma concentration of hyperforin. The extract with less hyperforin was inactive at the doses tested [29]. Similar results were published by Bhattacharya et al. [30]. They tested two *Hypericum* extracts of different hyperforin content (4.5% and 38.8%) in a battery of behavioural tests, two of them specific for antidepressant activity (FST and reserpine-syndrome in mice). In both models, they showed a strong correlation between the hyperforin concentration of the extract and efficacy. Another animal model in which clear responses to clinically active antidepressants are shown is the "Learned Helplessness" model described by Maier and Seligmann [31]. In this model, animals are exposed to stressors producing deficits in escape behaviour, "Learned Helplessness", which can be reversed by a wide range of clinically active antidepressants. Oral administration of hyperforin acetate at a dose of 10 mg/kg significantly reversed the number of escape failures in a manner comparable to the tricyclic antidepressant imipramine [32]. Treatment for seven consecutive days was necessary to elicit this antidepressant-like effect. In the forced swimming test, triple administration of this hyperforin salt was effective over a dose range of 5–20 mg/kg [32]. Experiments conducted with ethanolic or CO_2-extracts containing 4.5% or 38.8% hyperforin, showed a strong correlation between hyperforin content of the extract and efficacy in the "Learned Helplessness" and "FST" model in rats (Tab. 1) [16]. The influence of hyperforin on the hypothalamic–pituitary–adrenal axis (HPA) was investigated by Franklin et al. [33]. They used a supercritical CO_2 extract containing 31% hyperforin and no other known active constituents, according to HPLC analysis. Intraperitoneal application of the CO_2-extract resulted in a significant increase of plasma corticosterone level and a decrease of plasma prolactin level. The effects on corticosterone plasma concentration were prevented by co-administration of ketanserin, indicating that these effects are mediated by serotonin 2 receptors [33, 34]. In other experiments conducted in rats, the same authors showed that hyperforin attenuated the increase in plasma prolactin levels induced by haloperidol, and they speculated that hyperforin modulates dopaminergic functions in the brain [33]. Recently, it was shown that hyperforin at doses of 4 and 8 mg/kg significantly reduced the immobility time in the mouse "Tail Suspension Test", whereas it was inactive at doses of 2 and 20 mg/kg [35]. In further experiments, the same authors could show that step by step removal of either hyperforin or hypericin did not result in a loss of pharmacological activity. Such results clearly indicate that, besides hyperforin, other constituents contribute to the antidepressant activity of *Hypericum perforatum* extracts [35]. The bioavailability of hyperforin was evaluated preclini-

Table 1. Effects of ethanolic and CO_2-extracts of *Hypericum perforatum* and imipramine in the FST and the learned helplessness paradigm in rats [16]

Group (route of administration)	Dose (mg/kg/day)	FST Immobility (s)	Learned helplessness	
			EF	AR
Vehicle (p.o.)	–	165.4 ± 3.05	20.6 ± 0.63	1.6 ± 0.22
Ethanolic extract (p.o.)	50	149.6 ± 3.11**	17.5 ± 0.90**	4.0 ± 0.26***
	150	133.7 ± 1.91***	15.0 ± 0.65***	4.5 ± 0.32***
	300	115.4 ± 2.60***	10.6 ± 0.68***	5.2 ± 0.24***
CO_2-extract (p.o.)	5	153.9 ± 2.43*	19.2 ± 0.75	3.5 ± 0.32***
	15	131.1 ± 3.06***	14.7 ± 0.75***	4.9 ± 0.29***
	30	119.43 ± 3.01***	10.9 ± 0.69***	5.2 ± 0.24***
Imipramine (i.p.)	10	103.7 ± 2.55***	9.9 ± 0.54***	5.2 ± 0.26***

Number of escape failures (EF); avoidance responses (AR).
Statistically significant differences from the corresponding vehicle-treated groups are marked with *, **, or ***, representing $p < 0.05$; < 0.01 or < 0.001, respectively. All values given in the table are mean ± SEM.

cally in rats and mice. A plasma concentration of up to 400 ng/ml was measured in rats after single oral administration of 300 mg/kg ethanolic extract (5% hyperforin). In this study, a hyperforin concentration of more than 100 ng/ml persisted for about 10 h. These concentrations are comparable to those measured in human plasma after treatment with 1200 mg extract per day [36]. Similar results, after oral administration of *Hypericum* extract or hyperforin DCHA salt in male CD1-mice, were published by Cantoni et al. [37]. They found a plasma concentration of up to 100 ng/ml after single administration of 300 mg/kg hydroalcoholic extract (4.5% hyperforin) or 18.1 mg/kg of hyperforin DCHA salt. In mice, it was recently demonstrated that hyperforin reached the brain in sufficient amounts to produce a pharmacological effect after oral administration of 300 mg/kg ethanol extract or 15 mg/kg pure hyperforin-Na salt [38]. Hyperforin is therefore not only active as a non-specific neurotransmitter uptake inhibitor, but also shows antidepressant effects in different animal models indicative of clinical efficacy. To date, hyperforin is the only biologically active constituent of *Hypericum perforatum* known to pass the blood–brain barrier in concentrations which could explain the antidepressant activity of the extract.

Flavonol glycosides and biflavones

Different flavonol glycosides, with quercetin as the aglycone, represent a major group of biologically active compounds in *Hypericum perforatum*. In total, they represent up to 4% of the crude plant material [1, 39]. Some of them show *in vitro* and/or *in vivo* activities, indicating that they might contribute to the antidepressant activity of the extract.

In vitro effects of flavonols and biflavones

Spasmolytic *in vitro* activities of flavonols have been investigated using smooth muscle preparations from different animal species. MAO-A-inhibition was tested in isolated rat liver-mitochondria preparations and radioligand binding assays were performed using recombinant human or rat receptors or transporters.

Calcium-antagonistic effects of quercetin were studied in rat aortic strips, where quercetin inhibited the Ca^{2+}-induced contractions at concentrations between 10^{-5} M and 10^{-4} M [40]. Noradrenaline-induced contractions in guinea pig main pulmonary artery preparations were also inhibited by preincubation with quercetin [41]. It appears that the antispasmodic activity depends on the water solubility of the substances [42]. Trute et al. studied different flavone aglycones, and mono- or diglycosides in guinea pig ileum preparations with acetylcholine-induced contractions, and found that the aglycones were most active, followed by the corresponding mono- and diglycosides, the rutosides [42]. Similar results were found for inhibition of the enzyme monoamine oxidase type A (MAO-A), a main mechanism of antidepressant activity. Studies conducted by Sparenberg et al. showed most activity for the flavonoid-aglyca quercetin, kämpferol and luteolin, whereas the glycosides were less active [43]. The biflavones amentoflavone and biapigenine were inactive in this study. Inhibition of benzodiazepine binding by amentoflavone was reported by Baureithel et al. [44]. In this study, the biflavone I3, II8-biapigenin and the flavonoids rutin, hyperoside and quercitrin did not inhibit benzodiazepine binding up to concentrations of 1 µM. Recently, Butterweck et al. tested different constituents of St. John's Wort extract for their binding capacity to 42 different biogenic amine receptors or transporters [23]. The biflavonoid amentoflavone, which significantly inhibited binding at 5-HT_{1D}, 5-HT_{2C}, dopamine D_3, δ-opiate and benzodiazepine receptors, was most active in this study. Other flavones showed only weak activity. Quercetin inhibited binding to dopamine D_4-receptors, rutin to adrenergic α_{2C}-receptors, hyperoside to adrenergic α_{2A}-receptors and miquelianin to adrenergic α_{2C}, cholinergic M_2 and M_5 receptors [23].

In vivo effects of flavonols and biflavones

Antidepressant-like *in vivo* activities have been described for a number of flavonols and biflavones. Initial results with isolated flavonols in depression-relevant animal models were reported by Butterweck et al. in 2000 [6]. They

Table 2. Effects of different constituents in the FST [7]

Constituents	Dose mg/kg p.o.	Inhibition (% *versus* control)
Extract	300	59 *
Hyperforin-Na^+	30	66 *
Hyperosid	10	40 *
Isoquercitrin	0.6	21 *
I 3,II 8-Biapigenin	0.1	43 *
Quercetin	1.0	11
Quercitrin	1.0	2
Hypericin/Pseudohypericin	0.5	2
Rutin	10	0

Statistically significant differences from the corresponding vehicle-treated groups are marked with *, representing $p < 0.05$.

found that the flavones hyperoside, miquelianin and isoquercitirin significantly reduced the immobility time in the FST in rats [6]. These results were corroborated by Nöldner, who found that beside hyperoside and isoquercitirin, I3,II8-biapigenin reduced the immobility time in a modified version of the FST in rats [7].

Interestingly, some of the tested flavonoids showed activity over only a small dose range. This in part resulted in U-shaped dose-response curves [6].

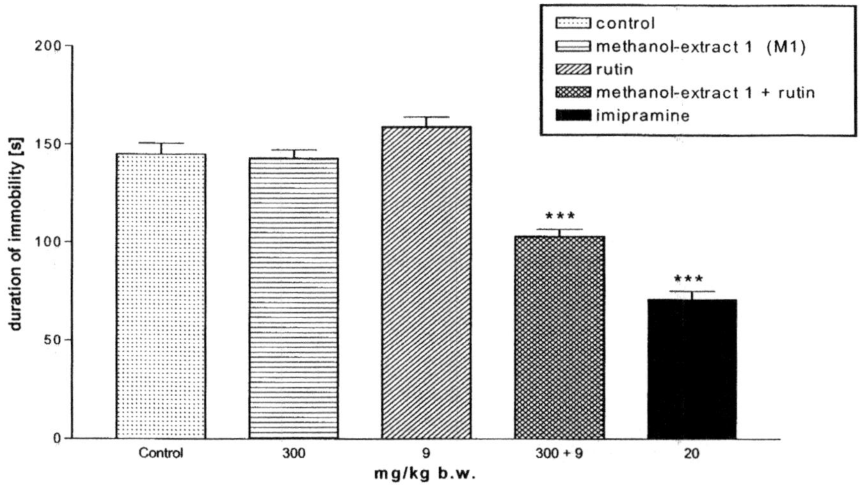

Figure 1. Effects of *Hypericum* extract with low rutin content, pure rutin and *Hypericum* extract with addition of rutin in comparison to imipramine in the FST [8]. ***, represents $p < 0.001$.

In addition to the intrinsic activities of flavones or biflavonoids, some activity-modifying effects were also reported recently. It was shown that rutin is necessary for the activity of *Hypericum* extracts in the FST in rats [8]. *Hypericum* extracts without rutin, or with very low rutin content, are completely inactive, while extracts containing sufficient amounts of rutin are always active in this test model [8]. The inactive extracts with low rutin content become active after adding rutin, indicating that rutin is necessary for this activity. It is known that the range in concentration of flavones in the crude plant varies widely depending on the developmental stage of the plant at the time of harvesting [45, 46]. In addition to this normal biological variation, it has been reported that there is a rutin-free chemotype that is morphologically indistinguishable from *Hypericum* species containing rutin [47]. *Hypericum* extracts prepared from this chemotype are inactive in the FST model [8].

Naphthodianthrones

Naphthodianthrones, i.e., hypericin and pseudohypericin, represent about 0.03–0.3% of the crude *Hypericum* plant material [39]. They are located in the flowering parts of the plant and aroused the early interest of phytochemists because of their intense red colour and their phototoxic properties [39, 48]. From a pharmacological point of view, hypericins are very interesting compounds. Apart from their phototoxic and antiviral activities, they have some pharmacological effects that could contribute to the antidepressant activity of the *Hypericum* extract.

In vitro effects of hypericin and pseudohypericin

Hypericin inhibits the nuclear transcription factor (NF-κB) over the micromolar range in HeLa and TC10 cells, indicating an anti-inflammatory activity [49]. However, publications regarding the effects of hypericin on MAO-activity appear to be contradictory. Suzuki et al. found an irreversible inhibition of MAO type A and B in rat brain mitochondria preparations, where the inhibition of type A was higher than that of type B [50]. Bladt and Wagner investigated hypericin and other *Hypericum* constituents in rat brain homogenates and did not find any MAO-inhibiting activity for pure hypericin [51]. Raffa investigated the affinity of hypericin at 30 receptor or uptake sites [52]. He found that hypericin had only moderate affinity for muscarinic cholinergic receptors and similar affinity for σ receptors (about 50% inhibition by 1 μM). Simmen et al. published that hypericin and pseudohypericin over the micromolar range showed affinity for different opioid receptors as well as for the CRF_1-receptor [53]. In a recently presented paper, the CRF-binding of hypericin and pseudohypericin was investigated in more detail [54]. The author measured the CRF-stimulated cAMP formation in recombinant Chinese hamster ovary (CHO) cells and found that pseudohypericin selectively antagonised the CRF_1-receptor (K_B 0.76 μM). In binding studies conducted on recombinant monoamine receptors, hypericin and pseudohypericin showed affinity for dopamine D_3 and D_4 receptors and hypericin for adrenergic β_1 and β_2-receptors in addition [23].

In vivo effects of hypericin and pseudohypericin

The potential contribution of hypericin or pseudohypericin to the antidepressant effects of St. John's Wort extracts has been investigated by Butterweck et al. In experiments using the rat FST they showed that the solubility of hypericin was increased in the presence of a procyanidin fraction. This significantly increased the *in vivo* effects [9]. They showed that hypericin and pseudohypericin are inactive during acute treatment, but become active on co-administration of a procyanidin fraction. This enhanced efficacy correlated to increased plasma levels of hypericin [55]. In a more recent publication, it was shown that after treatment over 14 days with pure hypericin at a dose of 0.1 mg/kg, the immobility time in the rat FST was reduced significantly [56]. Interestingly, in these experiments there was no difference between animals receiving pure hypericin and those receiving hypericin in combination with procyanidin. Although hypericin is active itself, removal of this constituent from the St. John's Wort extract did not result in loss of efficacy [35]. Chronic treatment over 14 days reduced plasma ACTH and corticosterone levels significantly, suggesting a decrease in the functional activity of the HPA axis [56]. In a more detailed study, the influence of hypericin on the expression of genes that may be involved in the regulation of the HPA axis was investigated using the *in situ* hybridisation histochemistry technique [10]. Chronic treatment for 8 weeks significantly decreased levels of CRF mRNA by 16–22% in the hypothalamus, and serotonin $5\text{-}HT_{1A}$ receptor mRNA by 11–17% in the hippocampus [10].

Concluding remarks

St. John's Wort extracts, like other herbal remedies, are complex mixtures of many different constituents. Only a few of the constituents have been identified and isolated, enabling pharmacological studies to be performed. The findings discussed here demonstrate that the pharmacological effects of the single constituents differ when given alone or in combination with other constituents, indicating that the extract is more than the sum of the single compounds. Some constituents, which were not active when tested alone, may act synergistically and help other constituents to become effective (e.g., rutin or the procyanidins). The data published so far strongly indicate that at least hyperforin, hypericin and some flavones and biflavonoids contribute to the antidepressant activity of *Hypericum perforatum* preparations. Only extracts which contain these constituents in sufficient amounts show antidepressant activity.

References

1 Butterweck V (2003) Mechanism of action of St John's Wort in depression. *CNS Drugs* 17: 539–562

2 Müller WE, Singer A, Wonnemann M, Hafner U, Rolli M, Schäfer C (1998) Hyperforin represents the neurotransmitter reuptake inhibiting constituent of *Hypericum* extract. *Pharmacopsychiatry* 31 (Suppl): 16–21

3 Kaehler ST, Sinner C, Chatterjee SS, Philippu A (1999) Hyperforin enhances the extracellular concentrations of catecholamines, serotonin and glutamate in the rat locus coeruleus. *Neurosci Lett* 262: 199–202

4 Müller WE, Singer A, Wonnemann M (2001) Hyperforin – Antidepressant activity by a novel mechanism of action. *Pharmacopsychiatry* 34 (Suppl 1): 98–102

5 Singer A, Wonnemann M, Müller WE (1999) Hyperforin, a major antidepressant constituent of St. John's Wort, inhibits serotonin uptake by elevating free intracellular Na^+. *J Pharmacol Exp Ther* 290: 1363–1368

6 Butterweck V, Jürgenliemk G, Nahrstedt A, Winterhoff H (1999) Flavonoids from *Hypericum perforatum* show antidepressant activity in the Forced Swimming Test. *Planta Med* 66: 3–6

7 Nöldner M (2001) Antidepressive effekte von johanniskraut und wichtigen inhaltsstoffen im tiermodell. In: WE Müller, M Gastpar (eds): *Behandlung von depressiven Patienten in der täglichen Praxis; Stellenwert von Johanniskrautextrakt.* LinguaMed Verlags GmbH, 31–42

8 Nöldner M, Schötz K (2002) Rutin is essential for the antidepressant activity of *Hypericum perforatum* extracts in the Forced Swimming Test. *Planta Med* 68: 577–580

9 Butterweck V, Petereit F, Winterhoff H, Nahrstedt A (1998) Solubilized hypericin and pseudohypericin from *Hypericum perforatum* exert antidepressant activity in the Forced Swimming Test. *Planta Med* 64: 291–294

10 Butterweck V, Winterhoff H, Herkenham M (2001) St John's Wort, hypericin, and imipramine: a comparative analysis of mRNA levels in brain areas involved in HPA axis control following short-term and long-term administration in normal and stressed rats. *Mol Psychiatry* 6: 547–564

11 Gurevich AI, Dobrynin VN, Kolosov MN, Popravko SA, Ryabova ID, Chernov BK, Derbentseva NA, Aizenman BE, Gargulya AD (1971) Hyperforin an antibiotic from *Hypericum perforatum. Antibiotiki* 16: 510–513

12 Hübner AT (2003) Treatment with *Hypericum perforatum* L. does not trigger decreased resistance in *Staphylococcus aureus* against antibiotics and hyperforin. *Phytomedicine* 10: 206–208

13 Schempp CM, Pelz K, Wittmer A, Schopf E, Simon JC (1999) Antibacterial activity of hyperforin from St. John's Wort against multiresistant *Staphylococcus aureus* and gram-positive bacteria. *Lancet* 353: 2129

14 Chatterjee SS, Nöldner M, Koch E, Erdelmeier C (1998) Antidepressant activity of *Hypericum perforatum* and hyperforin: the neglected possibility. *Pharmacopsychiatry* 31 (Suppl): 7–15

15 Eckert GP, Müller WE (2001) Effects of hyperforin on the fluidity of brain membranes. *Pharmacopsychiatry* 34 (Suppl 1): 22–25

16 Chatterjee SS, Bhattacharya SK, Wonnemann M, Singer A, Müller WE (1998) Hyperforin as a possible antidepressant component of *Hypericum* Extracts. *Life Sci* 63: 499–510

17 Uebelhack R, Franke L (2001) *In vitro* effects of St. John's Wort extract and hyperforin on 5HT uptake and efflux in human blood platelets. *Pharmacopsychiatry* 34 (Suppl 1): 146–147

18 Fisyunov AI, Lozovaya NA, Tsyntsadze TSh, Yatsenko NM, Chatterjee SS, Krishtal OA (2001) Modulating action of hyperforin on the P-type calcium channels in the membranes of rat cerebellar Purkinje neurons. *Neurophysiology* 33: 8–14

19 Krishtal O, Lozovaya N, Fisunov A, Tsintsadze T, Pankratov Y, Kopanitsa M, Chatterjee SS (2001) Modulation of ion channels in rat neurons by the constituents of *Hypericum perforatum. Pharmacopsychiatry* 34 (Suppl 1): 74–82

20 Fisunov A, Lozovaya N, Tsintsadze T, Chatterjee S, Nöldner M, Krishtal O (2000) Hyperforin modulates gating of P-type Ca^{2+} current in cerebellar Purkinje neurons. *Pflügers Arch – Eur J Physiol* 440: 427–434

21 Chatterjee S, Filippov V, Lishko P, Maximyuk O, Nöldner M, Krishtal O (1999) Hyperforin attenuates various ionic conductance mechanisms in the isolated hippocampal neurons of rat. *Life Sci* 65: 2395–2405
22 Simmen U, Burkard W, Berger K, Schaffner W, Lundstrom K (1999) Extracts and constituents of *Hypericum perforatum* inhibit the binding of various ligands to recombinant receptors expressed with the Semliki Forest Virus system. *J Recept Signal Transduct Res* 19: 59–74
23 Butterweck V, Nahrstedt A, Evans J, Hufeisen S, Rauser L, Savage J, Popadak B, Ernsberger P, Roth BL (2002) *In vitro* receptor screening of pure constituents of St. John's Wort reveals novel interactions with a number of GPCRs. *Psychopharmacology* 162: 193–202
24 Saxena PR, Lawang A (1985) A comparison of cardiovascular and smooth muscle effects of 5-hydroxytryptamine and 5-carboxamidotryptamine, a selective agonist of 5-HT_1 receptors. *Arch Int Pharmacodyn* 277: 235–252
25 Greenshaw AJ (1993) Behavioral pharmacology of 5-HT_3 receptor antagonists: a critical update on therapeutic potentials. *TIPS* 14: 265–270
26 Porsolt RD, Anton G, Blavet N, Jaffre M (1978) Behavioral despair in rats: a new model sensitive to antidepressant treatments. *Eur J Pharmacol* 47: 379–391
27 Porsolt RD (1981) Behavioural despair. In: SJ Enna, JB Malick, E Richelson (eds): *Antidepressants: neurochemical, behavioural and clinical perspectives*. Raven Press, New York, 121–139
28 Klusa V, Germane S, Nöldner M, Chatterjee SS (2001) *Hypericum* extract and hyperforin: Memory-enhancing properties in rodents. *Pharmacopsychiatry* 34 (Suppl 1): 61–69
29 Cervo L, Rozio M, Ekalle-Soppo CB, Guiso G, Morazzoni P, Caccia S (2002) Role of hyperforin in the antidepressant-like activity of *Hypericum perforatum* extracts. *Psychopharmacology* 164: 423–428
30 Bhattacharya SK, Chakrabarti A, Chatterjee SS (1998) Activity profiles of two hyperforin-containing *Hypericum* extracts in behavioral models. *Pharmacopsychiatry* 31 (Suppl): 22–29
31 Maier SF, Seligman MEP (1976) Learned helplessness: theory and evidence. *J Exp Psychol* 1: 3–46
32 Zanoli P, Rivasi M, Baraldi C, Baraldi M (2002) Pharmacological activity of hyperforin acetate in rats. *Behav Pharmacol* 13: 645–651
33 Franklin M, Chi JD, Mannel M, Cowen PJ (2000) Acute effects of LI 160 (extract of *Hypericum perforatum*, St. John's Wort) and two of its constituents on neuroendocrine responses in the rat. *J Psychopharmacol* 14: 360–363
34 Franklin M, Cowen PJ (2001) Researching the antidepressant actions of *Hypericum perforatum* (St. John's Wort) in animals and man. *Pharmacopsychiatry* 34 (Suppl 1): 29–37
35 Butterweck V, Christoffel V, Nahrstedt A, Petereit F, Spengler B, Winterhoff H (2003) Step by step removal of hyperforin and hypericin: activity profile of different *Hypericum* preparations in behavioral models. *Life Sci* 73: 627–639
36 Biber A, Fischer H, Römer A, Chatterjee SS (1998) Oral bioavailability of hyperforin from *Hypericum* extracts in rats and human volunteers. *Pharmacopsychiatry* 31 (Suppl): 36–43
37 Cantoni L, Rozio M, Mangolini A, Hauri L, Caccia S (2003) Hyperforin contributes to the hepatic CYP3A-inducing effect of *Hypericum perforatum* extract in the mouse. *Toxicol Sci* 75: 25–30
38 Keller JH, Karas M, Müller WE, Volmer DA, Eckert GP, Tawab MA, Blume HH, Dingermann T, Schubert-Zsilavecz M (2003) Determination of hyperforin in mouse brain by high-performance liquid chromatography/tandem mass spectrometry. *Anal Chem* 75: 6084–6088
39 Nahrstedt A, Butterweck V (1997) Biologically active and other chemical constituents of the herb of *Hypericum perforatum* L. *Pharmacopsychiatry* 30 (Suppl): 129–134
40 Morales MA, Lozoya X (1994) Calcium-antagonist effects of quercetin on aortic smooth muscle. *Planta Med* 60: 313–317
41 Abdalla S, Zarga MA, Afifi F, Al-Khalil S, Mahasneh A, Sabri S (1989) Effects of 3,3'-di-O-methylquercetin on guinea-pig isolated smooth muscle. *J Pharm Pharmacol* 41: 138–141
42 Trute A, Gross J, Mutschler E, Nahrstedt A (1997) *In vitro* antispasmodic compounds of the dry extract obtained from *Hedera helix*. *Planta Med* 63: 125–129
43 Sparenberg B, Demisch L, Hölzl J (1993) Untersuchungen über antidepressive Wirkstoffe von Johanniskraut. *Pharm Ztg Wiss* Nr.2 6./138: 50–54
44 Baureithel KH, Berger Büter K, Engesser A, Burkard W, Schaffner W (1997) Inhibition of benzodiazepine binding *in vitro* by amentoflavone, a constituent of various species of *Hypericum*. *Pharm Acta Helv* 72: 153–157

45 Tekel'ová D, Repcák M, Zemková E, Tóth J (2000) Quantitative changes of dianthrones, hyperforin and flavonoids content in the flower ontogenesis of *Hypericum perforatum*. *Planta Med* 66: 778–780
46 Umek A, Kreft S, Kartnig T, Heydel B (1999) Quantitative phytochemical analyses of six *Hypericum* species growing in Slovenia. *Planta Med* 65: 388–390
47 Mártonfi P, Repcák M, Ciccarelli D, Garbari F (2001) *Hypericum perforatum* L. – Chemotype without rutin from Italy. *Biochem Syst Ecol* 29: 659–661
48 Greeson JM, Sanford B, Monti DA (2001) St. John's Wort (*Hypericum perforatum*): a review of the current pharmacological, toxicological, and clinical literature. *Psychopharmacology* 153: 402–414
49 Bork PM, Bacher S, Schmitz ML, Kaspers U, Heinrich M (1999) Hypericin as a non-antioxidant inhibitor of NF-κB. *Planta Med* 65: 297–300
50 Suzuki O, Katsumata Y, Oya M, Bladt S, Wagner H (1984) Inhibition of monoamine oxidase by hypericin. *Planta Med* 50: 272–274
51 Bladt S, Wagner H (1993) MAO-Hemmung durch Fraktionen und Inhaltsstoffe von *Hypericum*-Extrakt. *Nervenheilkunde* 12: 349–352
52 Raffa RB (1998) Screen of receptor and uptake-site activity of hypericin component of St. John's Wort reveals σ receptor binding. *Life Sci* 62: 265–270
53 Simmen U, Higelin J, Berger-Büter K, Schaffner W, Lundstrom K (2001) Neurochemical studies with St. John's Wort *in vitro*. *Pharmacopsychiatry* 34 (Suppl 1): 137–142
54 Simmen U, Bobirnac I, Ullmer C, Lübbert H, Berger-Büter K, Schaffner W, Schoeffter Philippe (2003) Antagonist effect of pseudohypericin at CRF_1 receptors. *Eur J Pharmacol* 458: 251–256
55 Butterweck V, Liefländer-Wulf U, Winterhoff H, Nahrstedt A (2003) Plasma levels of hypericin in presence of procyanidin B2 and hyperoside: A pharmacokinetic study in rats. *Planta Med* 69: 189–192
56 Butterweck V, Korte B, Winterhoff H (2001) Pharmacological and endocrine effects of *Hypericum perforatum* and hypericin after repeated treatment. *Pharmacopsychiatry* 34 (Suppl 1): 2–7

Edited by W. E. Müller

Hyperforin and efficacy in animal models for anxiety and cognition

Shyam S. Chatterjee

Pharmacology Laboratories, Research and Development Department, Dr. Willmar Schwabe GmbH & Co KG, Willmar-Schwabe-Straße 4, 76227 Karlsruhe, Germany

Introduction

Although diverse medicinal uses of various types of St. John's Wort (SJW) preparations have been known for centuries, most modern reports on therapeutic potentials of this herb concentrate mainly on antidepressant-like activity of its hydroalcoholic extracts (SJWEs). One of the very first reports based on therapeutic observations suggesting usefulness of a SJWE for treatment of depressed patients, appeared much earlier [1] than the fortuitous discovery of analogous effects of imipramine or iproniazid. However, reports on controlled clinical trials with diverse types of SJWEs in patients with mild-to-moderately severe depressive disorders started appearing after the 1980s. Results of these earlier clinical trials, and other efforts to define antidepressant-like effects of SJWEs, and their bioactive components, are summarised in a supplement of *Pharmacopsychiatry* [2].

However, during the early 1990s, extensive efforts were made in our laboratories to evaluate other therapeutic potentials of SJW traditionally known for this herb. One major pharmacological observation during these efforts was that hyperforin, i.e., quantitatively the major and structurally a unique antimicrobial secondary metabolite of the herb, inhibits serotonin induced contractions of isolated guinea pig ileum [3]. Subsequent efforts made to evaluate *in vivo* relevance of this observation led to a speculative working hypothesis that hyperforin could be a functional antagonist of central serotoninergic 5-HT3 and/or 5-HT4 receptors [4]. Since specific antagonists of these receptors were known to possess antidepressant and anxiolytic-like effects in animal behavioural models [5, 6], potential effects of hyperforin and a few experimental SJW extracts were tested in such models. Results of these efforts, and numerous others made in different laboratories, have been summarised also in two recent supplements of *Pharmacopsychiatry* [7, 8].

In light of these reports, it could be concluded that hyperforin is quantitatively the major psychoactive component of SJW, and that depending on its concentration in SJWEs their pharmacological activity profiles can vary considerably. In addition, since along with its antidepressant-like activities hyperforin also revealed anxiolytic-like and cognitive function improving properties in rodent models, it was speculated that hyperforin-containing SJWEs could represent a novel type of psychotherapeutic. Subsequent reports describing anxiolytic and memory improving effects of SJWEs in animal models add further experimental evidences to this hypothesis. However, the role of hyperforin in these activities of SJWEs is not yet precisely defined and therapeutic relevance of these preclinical observations still remain uncertain. Available information potentially useful for better clarification of the situation will be summarised and critically analysed in this chapter. In addition, some possible therapeutic implications of these findings will be pointed out.

It must be mentioned though, that putative anxiolytic and/or antidepressant-like effects of SJWEs have been implicated in their alcohol intake reducing effects observed in rodents also [9–14]. Although the observations presented in one of these reports [12] strongly suggest that hyperforin is the major SJW alcohol consumption reducing component, another one [9] describes similar effects of a SJWE almost devoid of hyperforin. Since these observations and speculations have been critically reviewed very recently [14], they are not discussed here again.

Effects on anxiety

The conventionally known rat elevated plus-maze model for pharmacological screening and characterisation of anxiolytics was used to detect such potential effects of purified hyperforin. Although after a higher single oral dose (10 mg/kg), no significant effects of hyperforin were detected, after all the three lower doses tested (0.3, 1 and 3 mg/kg), its clear anxiolytic-like effect was apparent [4]. As indicated before, these experiments were planned as a consequence of observations suggesting functional antagonism of 5-HT3 or 5-HT4 receptor mediated responses by a single 10 mg/kg oral dose of purified hyperforine. However, since anxiolytic-like activity was observed after its lower doses only, and possible anti-serotoninergic effects of lower hyperforin doses have not yet been evaluated, involvement of these receptors in observed anxiolytic-like activity remains uncertain. Attempts to verify this possibility by *in vitro* ligand displacement experiments do not indicate any affinity of hyperforin for 5-HT3, 5-HT4 or other known serotonin receptors. As a matter of fact, despite extensive efforts [15–17, and unpublished observations], no specific hyperforin binding site potentially involved in its observed low dose anxiolytic-like activity, or in its other known behavioural effects, could yet be detected with certainty by such *in vitro* binding studies.

As described later, lower oral hyperforin doses (i.e., below 10 mg/kg) have revealed not only its anxiolytic-like activity in elevated plus-maze model, but also its cognitive function modifying activities in animal models [18]. However, such is not the case for its antidepressant-like efficacy in animal

Table 1. Reported active doses of hyperforin in rat models for anxiety, depression and cognitive functions

Model (for)	Effective doses of hyperforin
Elevated plus-maze (Anxiety)	Single oral doses between 0.3–3 mg/kg administered 1 h before test are almost equi-active, whereas after 10 mg/kg no significant effects are observed [4].
Behavioural despair (Depression)	Effective after repeated doses only, and efficacy enhances with the number of treatment days. After 9 daily treatments, the active dose range was 10–30 mg/kg/day. After this treatment schedule, no effects could be detected with 0.15, 1.5 or 100 mg/kg/day doses [18].
Escape deficit (Depression?)	Single i.p. doses between 25–75 mg/kg administered 30 min before unavoidable stress are almost equi-active, whereas after 12.5 mg/kg no significant effects are observed [19].
Conditioned–avoidance response (Cognitive effects)	Both daily oral doses tested (1.25 and 2.5 mg/kg) started facilitating learning after the second day of seven consecutive days of treatment. However, the lower dose effects were quantitatively more pronounced, and significant memory consolidation effects were observed after the lower one only [18].

models, where its activities are detected only after repeated daily doses higher than 10 mg/kg (see Tab. 1). These observations, taken together with the ones made with an experimental hyperforin enriched (38.8%) extract, have led to the suggestion that the primary mechanism triggered by hyperforin, and responsible for its observed anxiolytic-like activity, is prone to adaptive changes, and that these adaptive responses could be involved in its antidepressant-like activity [20]. It has been demonstrated indeed, that hyperforin induces adaptive changes in the functions of various neurotransmitter mediated processes, including some serotonin mediated ones, after its higher or repeated doses [4, 21]. Although these circumstantial evidences or speculative suggestions based on experimental observations do not yet allow definitive statements on possible involvement of serotonin mediated processes in observed low dose effects of hyperforin, they could be useful for clarifying the neurotransmitter systems involved in its reported low dose effects in animal models.

Soon after anxiolytic and antidepressant-like activities of purified hyperforin were identified, an attempt was made to clarify the role of hyperforin in the behavioural activity profiles of SJWEs [20]. In this study, effects of a therapeutically used SJWE in a battery of conventionally known rodent behavioural models were compared with those of a hyperforin enriched extract devoid of all other bioactive constituents of SJW identified by the mid 1990s. Three graded doses of each extract, corresponding to their inactive, low and high antidepressant ones in rat models were used in this study, and since three daily doses of the extracts were necessary for estimating these doses, this treatment schedule was used for comparison sake. One major difference between

Table 2. Behavioural activity profiles of hyperforin and of a therapeutically used SJWE (Neuroplant®) in rodent behavioural models

Activity	Hyperforin	SJWE
Antidepressant like [4, 18]	+	+
Anxiolytic like [4, 20]	+	-
Memory acquisition [18]	+	+
Memory consolidation [18]	+	+
Anti-amnestic (or cholinergic) [18]	+	*
5-HTP- potentiating [20]	+	+
Central dopamin-potentiating [20]	–	+

+ = present; – = absent; * = not yet properly tested.

their observed activity profiles was their effects in models used for detecting potential anxiolytics. Thus, in the elevated plus-maze test for anxiolytics, the two lower doses of both the extracts did not alter mice behaviour on the maze. However, after the higher antidepressant doses were tested, one of them revealed anxiolytic-like activity, whereas the observations made with the other one indicated its mild anxiogenic-like activity. Since the doses of hyperforin administered with the extracts were comparable, and were within its anxiolytic ones, the observed opposite effects of the two extracts were attributed to the presence of different types of other components in the two extracts. The observations that all dopaminergic responses, studied during this profiling effort, were dose dependently enhanced by the therapeutically used SJWE only, and no anxiolytic-like effect could be detected in this extract, are in agreement with this inference (see Tab. 2). It was concluded, therefore, that some as yet unidentified SJWE component with potentiating effects on central dopaminergic functions antagonises anxiolytic like activity of hyperforin.

However, putative anxiolytic-like activities of four other types of SJWEs in rodent behavioural models have consistently been inferred by the authors of all subsequent reports [22–27]. One of them [22] was a short communication dealing with the effects of four medicinal plant extracts in animal behavioural models. Although many details on methods and observations made during the study are missing in this communication, it was mentioned that an aqueous SJWE prepared from aerial parts of the herb growing in Portugal was tested 30 min after its intraperitoneal doses. Conclusion of the authors was that anxiolytic-like effect of the extract in mouse elevated plus-maze model is significant after its lowest dose tested, i.e., 5 mg/kg, and their speculative suggestion was that hyperforin is not involved in this effect of the tested SJWE.

In another report [23], activity profile of a 50% aqueous ethanolic extract from whole plant, designated as Indian *Hypericum perforatum*, in five well known rat models for anxiolytics is described. The extract (100 and 200 mg/kg/day; p.o.) was administered for three consecutive days, and its effects, observed 1 h after the last doses, were compared with those of a single lorazepam dose (0.5 mg/kg; i.p.). Dose dependent anxiolytic effects of the extract in open field, elevated plus-maze, and novelty induced feeding tests were less pronounced than those of the anxiolytic. In elevated zero-maze and social interaction tests, extract effects, observed after the higher dose only, were similar to those of the benzodiazepine. The authors conclude that the extract is a milder anxiolytic than lorazepam.

Effects of an Italian SJWE and purified hypericins on rat exploratory activities in open-field and light–dark box were compared in another study [24]. Single very high extract doses and corresponding ones of hypericins were orally administered 1 h before the tests. Although significant stimulating effects of the extract on the rearing activities in open-field were observed even after its lowest dose tested (926 mg/kg), simultaneously quantified parameters for horizontal activities were influenced by its highest dose (2,778 mg/kg!) only. In the light–dark box test, only the intermediate extract dose (1,852 mg/kg!) significantly increased latency time of animals for the first transitions from the light to the dark compartment. This effect of the extract was fully antagonised by the benzodiazepine receptor antagonist flumazenil. No anxiolytic-like effects of hypericins were detected in the tests used. However, in view of the exceptionally high doses used in this study, any potential therapeutic relevance of these observations remains doubtful.

The three more recent reports [25–27] dealt with the activity profile of a therapeutically used SJWE (LI 160) in rodent behavioural models useful for distinguishing the effects of agents in diverse types of anxiety. Its single or repeated oral doses (62.5–500 mg/kg) did not alter locomotion of rats or their responses to cat odour [25]. Anxiolytic-like effects of the extract in rat elevated T-maze was observed after an intermediate single oral dose (125 mg/kg) only, which disappeared after repeated daily administrations. After the highest repeated daily dose tested (250 mg/kg/day), the extract revealed clear anxiogenic-like effect in the T-maze paradigm, and also significantly raised the number of transitions in the light–dark box test. The authors concluded though, that the extract exerts anxiolytic-like effects in a specific subset of defensive behaviour, particularly those related to generalised anxiety.

In the second study [26], effects of the extract were compared with those of paroxetine (5 mg/kg) in the so-called Mouse Defence Test Battery (MDTB). The tests were conducted 1 h after one single or 7 or 21 daily extract doses (150 or 300 mg/kg) administered orally. Significant effects of the extract were observed in the groups treated for 21 consecutive days with the higher extract dose only. Observed activity spectrum of paroxetine in MDTB was, neither qualitatively nor quantitatively, like that of the extract. These observations suggest that anti-panic and/or anxiolytic-like effects of the tested SJWE appear after repeated doses only, and that its activity spectrum in this test battery are not like those of anxiolytics or of some antidepressants. Results of an analogous comparative study in rat elevated T-maze model, reported later by the same authors [27], are in agreement with such inferences. Results of this later study suggest further that repeated administrations of antidepressive extract doses induce also its observed anxiolytic-like and/or anti-panic effects in the paradigm used.

Taken together, the results of behavioural studies conducted to date reveal that the anxiolytic like activity of SJWEs can not be predicted by their contents of hyperforin only. They suggest further that this activity of any given SJWE depend not only on its dose, but also on treatment duration. In any case, neither hyperforin nor any of the SJWEs tested to date can be regarded as benzodiazepin like anxiolytics.

Cognitive effects

The report dealing with anxiolytic like activity of the Indian SJWE [23] appeared, almost simultaneously, with another one dealing with its effects in animal models of cognitive dysfunction [28]. Although these two reports do not mention any analytical details of the extract tested, a later report from the same laboratories [29] mentioned that it was standardised to contain 4.5–5% hyperforin. In both these later mentioned reports, the effects of 100 or 200 mg/kg extract administered orally to rats for three consecutive days before the training sessions are described. Although effects of the treatments in step-through passive avoidance tests used were equivocal, clear memory enhancing and antagonistic effects of the extract against scopolamine or sodium nitrite or electroshock-induced amnesia were observed in elevated plus-maze and in active avoidance paradigms. In most of these tests, the observed effects were not dose dependant, and in a few, the active extract doses were the higher ones only. All observed effects of the extract were reported to be persistent even after nine treatment free days.

Effects of low doses (2–25 mg/kg; i.p.) of a Chinese SJWE containing 3% hyperforin in mouse step-down passive avoidance paradigm are described in another report [30]. Dose dependant beneficial effects on memory retrieval were observed in this study after administering the extract 30 min before the retrieval test conducted 24 h after training sessions. However, the same doses of the extract failed to reverse amnesia induced by scopolamine (3 mg/kg; i.p.; 15 min before retrieval test). Results of interaction studies conducted by co-administering various well-known receptor antagonists led the author to conclude that adrenergic and serotonergic 5-HT1A receptors are involved in the memory retrieval improving effects of the extract.

Beneficial effects of daily low oral doses of a dried SJW herb sample (commercialised in Poland as Hyperherba®) on learning and memory of rats in Morris water maze has been described more recently [31]. The herb contained 0.3% hypericin, and with its tested doses 4.3 or 13 µg/kg/day of hypericin was administered. Behavioural testing was conducted on five consecutive days during the 9th week of treatment. Marked improvements in learning and consolidation of spatial memory were observed in the animals treated with the higher herb dose only.

Potential effects of the therapeutically used extract (used for manufacturing Neuroplant®), in rodent avoidance tests were evaluated during a study designed to compare the efficacy of hyperforin (as sodium salt) with that observed after its administration with the extract [18]. The daily oral doses of purified hyperforin tested in the conditioned avoidance paradigm in rats were 1.25 and 2.5 mg/kg, whereas those of the SJWE were 25, 50, 150 and 300 mg/kg (which represented 1.25, 2.5, 7.5 and 15 mg/kg doses of hyperforin), and the treatments were given 1 h before daily training sessions. In these experiments, the lowest tested hyperforin dose and 50 mg/kg dose of SJWE considerably improved learning ability of rats in conditioned avoidance paradigm from day 2 onwards until day 7 of the treatments. In addition, the memory of the learned responses, acquired during seven consecutive days of treatment and training, was largely retained even after 9 days without further training and drug free period. Similar effects of the intermediate (150 mg/kg) or the lowest (25 mg/kg) dose of SJWE, or of the higher one of purified hyperforin (2.5 mg/kg), were less pronounced. After the highest one of the extract (300 mg/kg), almost no significant treatment effects were observed.

In view of these observations, the effects of a low single oral dose (25 mg/kg) of the extract, and that of hyperforin (1.25 mg/kg) were compared in mouse step-through passive avoidance paradigm [18]. No significant effects of the extract on memory acquisition or on scopolamine-induced amnesia were detected, whereas improved acquisition and retention of the memory of the learning task were observed for hyperforin. In addition, scopolamine-induced amnesia was completely reversed by hyperforin. These observations led the authors to conclude that cognitive effects of hyperforin are modulated by other extract components, and that cognitive function improving activities of hyperforin, or of the tested extract, are independent of their antidepressant or anxiolytic-like activities observed in rodents. Another therapy relevant conclusion of this study was that, after repeated lower doses, hyperforin is a more potent memory function improving agent than an antidepressant.

It appears, therefore, that hyperforin is indeed the most potent cognitive function improving component of SJW, and that, like in the case of its antidepressant and anxiolytic-like activities, its memory function modulating effects are altered also by other bioactive components of the herb. In view of the findings that very low doses of hyperforin improves memory acquisition and consolidation, and antagonises amnesia induced by the anticholinergic scopolamine, reports dealing with its effects on cholinergic neurotransmission could be of potential mechanistic and therapeutic interest. Therefore, they will be summarised below.

Hyperforin and central cholinergic neurotransmission

The very first experimental observations indicating possible involvement of cholinergic neurotransmission in the mode of action of hyperforin were that it inhibits not only contraction of guinea pig ileum induced by serotonin, but also those triggered by acetylcholine, nicotine, and many other neurotransmitters known to cause contractions by releasing endogenous acetylcholine [4]. However, in view of the facts that despite extensive efforts, no cholinergic effects of hyperforin could be detected in numerous other pharmacological models used to identify it as the major neuroactive component of SJWEs [3, 4, and unpublished observations], and involvement of cholinergic neurotransmission in the antidepressant effects of drugs is not well recognised or defined, little efforts were made to clarify its observed effects on smooth muscles.

Later pharmacological screening efforts made to identify other neuroactive components of SJWEs [32] led again to the observation that hyperforin, in concentrations as low as 37 nM, inhibits not only neuronal currents gated by various amino acids and ATP, but also those gated by stimulation of cholinergic receptors (observed inhibition = 41.7%). These efforts revealed in addition that intra-neuronal application of even very high concentration of hyperforin has no measurable effects, whereas its extracellular sub-μM concentrations inhibit several voltage-gated ion channels also. Since it has repeatedly been pointed out that many effects of hyperforin, observed *in vitro*, are due to its ability to modulate cellular ionic homeostasis, its observed effects on cholinergic responses were considered to be a consequence of its actions on an extracellular site of neurones regulating their ionic concentrations. In addition, since many effects of hyperforin observed *in vitro* are antagonised by serum albumin (unpublished observations), and its brain concentrations were below the detection level of many sensitive analytical methods tried initially, not much attention was paid to its cholinergic function modulating potentials. It must be mentioned though, that a more sensitive analytical method has become available now [33]. Using this method, mean brain hyperforin concentrations detected 3 h after oral administration of 15 mg/kg hyperforin as sodium salt or with SJWE were 28.8 and 15.8 ng/g, respectively.

Several recent findings strongly suggested that hyperforin activates some specific sodium channels only, and that this effect of the agent might explain many, if not all, of its effects observed *in vitro* [34, 35]. Since neuronal synthesis of acetylcholine is controlled by sodium-dependent high affinity choline uptake, efforts were made to characterise its potential effects on choline uptake and acetylcholine release [36]. A more therapy relevant *in vivo* observation made during this study was that systemic low dose (1 mg/kg; i.p.) hyperforin administration actually stimulated acetylcholine release and decreased choline concentrations in corpus striatum of rats. Although no definite statements can yet be made on possible mechanisms involved in this observed effect, or on its relevance for behavioural activity profile of hyperforin, this report was the very first one revealing a lower dose effect of hyperforin on the functions of a neurotransmitter system in the brains of freely moving animals.

A subsequent report from the same group [37] demonstrates that the lower hyperforin dose also stimulates hippocampal acetylcholine release in freely moving rats, and that this effect persists for a prolonged period after its administration. Since this effect was almost completely suppressed by local perfusion with calcium free buffer or with the sodium channel blocker tetrodotoxin, it was concluded that hyperforin releases acetylcholine by indirect mechanism, which is calcium-dependent, and requires neuronal communication and cell firing. It is now well recognised that the activity of septo-hippocampal cholinergic fibers is closely associated with cognitive functions in various tasks of attention, memory and learning [38, 39]. Therefore, it seems possible that observed cognitive effects of hyperforin is due to its facilitating effects on hippocampal acetylcholine release only. As a consequence of these observations, it can be said now that neglect of possible involvement of cholinergic neurotransmission in the modes of actions of hyperforin and SJWEs is no longer justifiable.

Therapeutic implications

More recent pharmacological findings discussed or referred to in this Chapter, reveal that activity profiles of SJWEs and hyperforin are neither functionally nor mechanistically like those of any known antidepressant or other psychoactive drug. Some major therapeutically interesting neuropharmacological properties of hyperforin-containing SJWEs revealed to date are listed in Table 3. It is apparent from this table, that SJW is an interesting source for therapeutics, especially useful for mild-to-moderately depressed patients with anxiety and cognitive abnormalities as other comorbid mental health conditions. However, since qualitatively as well as quantitatively the activity profiles of studied SJWEs vary considerably, and do not strictly depend on their hyperforin contents only, definitive statements on the nature of extract optimally suited for

Table 3. Major pharmacological activities of SJWEs and hyperforin potentially useful for predicting psychotherapeutic potentials of the herb. In many cases representative references are mentioned only.

1)	Antidepressant like activities in rodent models [2, 4, 7, 8, 17, 43]*
2)	Facilitation of learning ability and memory consolidation in rodents [18, 28–31, 43]*
3)	Anxiolytic and/or anxiogenic effects in rodents [4, 22–27]*
4)	Reduction of alcohol intake by rodents [9–14]*
5)	Efficacy against nicotine withdrawal [40]
6)	Neurotransmitter uptake inhibition and/or release stimulation: *in vitro* [2, 4, 7, 8,21, 43]
7)	Antagonism of diverse ion-channels and neurotransmitter functions: *in vitro* [4, 32, 43]
8)	Modulation of intracellular ionic homeostasis: *in vitro* [7, 8, 32, 34, 35, 43]
9)	Elevation of brain extra-cellular concentrations of various neurotransmitters in rats [8, 43]
10)	Stimulation of acetyl choline release in rat corpus striatum and hippocampus [36, 37]**
11)	Modulation of amyloid processing: *in vitro* [41–43]**

* Activities of SJWEs not necessarily dependent on their hyperforin contents only.
** Activities reported for pure hyperforin only.

such purposes is not yet possible. Observations that antidepressant and cognitive function improving doses of purified hyperforin do not overlap, and its anxiolytic-like activity diminishes after repeated or higher doses, indicate that it cannot be a therapeutic alternative for an *optimally standardised* SJWE. Pure hyperforin seems to be a novel type of potential antidementia agent with acetylcholine releasing activities and modulating effects on amyloid-processing [41–43]. Therefore, it could be a lead for the development of a potential therapy for cognitive abnormalities or for Alzheimer's disease and vascular dementia.

Until now, among diverse types of SJWEs used for preclinical studies, the two present in the most popular pharmaceuticals, i.e., Neuroplant® and Jarsin®, are the best pharmacologically defined ones. However, sufficient data enabling comparison of their antidepressant activities with their potential anxiolytic or cognitive function modulating effects in animal models are not yet available. They are manufactured and standardised to obtain SJWEs with optimal antidepressant-like activities, and it is fortuitous that other therapeutically interesting pharmacologically properties coexist in them. Therefore, it is not yet certain whether they are the most suitable ones for simultaneously obtaining all three therapeutic benefits offered by the herb. However, in view of the fact that both of them contain, until now, all known bioactive components of the herb, it is possible that some of their therapeutic benefits observed by patients, or by their physicians, are due to their effects on cognitive function or on anxiety also.

Cognitive disorders, anxiety and depression are the three major comorbid affective conditions encountered in a vast number of patients, and currently available synthetic psychotherapeutics, developed for patients with specific mental health problems, does not properly meet therapeutic demands of a vast number of these patients. Unfortunately, most psychotherapeutic drug development projects, including those using SJW as a starting material, continue to concentrate in the search of better, safer and more potent and specific leads, hits and potential drugs for each of these three disorders. This is because our current knowledge on aetiology, pathogenesis and progression of such comorbid conditions is not precise enough for rationally developing, identifying or characterising potential psychotherapeutics useful for combating the complications inherent in multiple ongoing mental health conditions. In view of the situation, the observations suggesting that an *optimally standardised* SJWE can offer therapeutic alternatives for simultaneous treatments of all the three major mental health problems, are of major therapeutic, scientific and economic interest.

In addition, precise knowledge of neurochemical mechanisms involved in the observed effects of hyperforin and other extractable bioactive SJW components in animal behavioural models could lead to novel pharmacological tools urgently needed for psychotherapeutic drug development purposes. Therefore, testing the validity of pharmacologically reasoned therapeutic predictions on the herb in patients with these three comorbid mental health conditions seems to be an urgent necessity. Results of numerous clinical trials with randomly selected SJWEs in patients with depressive symptoms only, and concentration of efforts to more precisely define their antidepressant-like pharmacological activities, do not seem to be suitable means for judging the most appropriate therapeutic potential of the herb.

Conclusions

Efforts to properly define therapeutic potentials of SJW must not be limited to studies with one of its diverse therapeutic potentials, or with one of its isolated bioactive secondary metabolites only. Although revelation of neuropharmacological properties of hyperforin have been helpful to precisely formulate a few therapy-relevant questions on the herb, it has not yet properly solved any of them. Similarly, concentration of efforts to characterise antidepressant-like therapeutic potentials of SJWEs for decades, has continued to deprive us of the knowledge of other psychotherapeutic potentials of the herb. SJW seems to be a definitive source for therapeutically interesting psychoactive molecules with novel modes of actions, and extracts containing proper combination of them could help achieve a therapeutic goal commonly neglected by almost all drug development projects. Pharmacologically, all studied SJWEs, as well as hyperforin, are polyvalent agents and are not like conventionally known antidepressants or other psychotherapeutics. Therefore, the most appropriate therapeutic potentials of hyperforin, or of SJWEs, cannot be judged, or defined, by a single bioassay, or by analytically defining hyperforin contents of the extracts only. At present, it can be concluded that neglect of the possibility that cholinergic neurotransmission is involved in some hyperforine-dependant psychotherapeutic potentials of the herb, and concentration of efforts to more precisely define antidepressant-like therapeutic potentials of SJWEs, have deprived us of a more rational therapy urgently needed for numerous mentally ill patients.

References

1 Daniel K (1935) Johanniskraut (*Hypericum perforatum*) bei psychischen Störungen. *Hippokrates* 14: 929–932
2 Müller WE, Kasper S (eds) (1997) A supplement of the journal. *Pharmarmacopsychiatry* 30 (Suppl 2): 71–134
3 Chatterjee SS, Koch E, Nöldner M, Biber A, Erdelmeier C (1996) Hyperforin and *Hypericum* extracts: Interaction with some neurotransmitter systems. *Phytomedicine* 3 (Suppl 1): 106
4 Chatterjee SS, Nöldner M, Koch E, Erdelmeier C (1998) Antidepressant activity of *Hypericum perforatum* and hyperforin: the neglected possibility. *Pharmacopsychiatry* 31 (Suppl 1): 7–15
5 Greenshaw AJ (1993) Behavioural pharmacology of 5-HT3 receptor antagonists: A critical update on therapeutic potentials. *TIPS* 14: 265–270
6 Eglen GR, Wong EHF, Dumuis A, Bockaert (1995) Central 5-HT4 receptors. *TIPS* 16: 391–398
7 Müller WE, Chatterjee SS (eds) (1998) Hyperforin and the antidepressant activity of St John's Wort. *Pharmacopsychiatry* 31 (Suppl 1): 1–60

8 Müller WE, Chatterjee SS (eds): (2001) Biocentre symposium on drug therapy: Advances of St. John's Wort research. *Pharmacopsychiatry* 34 (Suppl 1): 1–156
9 De Vry J, Maurel S, Schreiber R, de Beun R, Jentzsch KR (1999) Comparison of *Hypericum* extracts with imipramine and fluoxetine in animal models of depression and alcoholism. *Eur Neuropsychopharmacol* 9: 461–468
10 Panocka I, Perfumi M, Angeletti S, Ciccocioppo R, Massi M (2000) Effects of *Hypericum perforatum* extract on alcohol intake and on behavioral despair: A search for the neurochemical systems involved. *Pharmacol Biochem Behav* 66: 105–111
11 Perfumi M, Ciccocioppo R, Angeletti S, Cucculelli M, Massi M (1999) Effects of *Hypericum perforatum* extract on alcohol intake in Marchigian Sardinian alcohol-preferring rats. *Alcohol Alcohol* 34: 690–698
12 Perfumi M, Panocka I, Ciccocioppo R, Vitali D, Froldi R, Massi M (2001) Effects of a methanolic extract and hyperforin-enriched CO_2 extract of *Hypericum perforatum* on alcohol intake in rats. *Alcohol Alcohol* 36: 199–206
13 Rezvani AH, Overstreet DH, Yang Y, Clark E Jr, (1999) Attenuation of alcohol intake by extract of *Hypericum perforatum* (St John's Wort) in two different strains of alcohol-preferring rats. *Alcohol Alcohol* 34: 699–705
14 Rezvani AH, Overstreet DH, Perfumi M, Massi M (2003) Plant derivatives in the treatment of alcohol dependency. *Pharmacol Biochem Behav* 75: 593–606
15 Simmen U, Burkard W, Berger K, Schaffner W, Lundstrom K (1999) Extracts and constituents of *Hypericum perforatum* inhibit the binding of various ligands to recombinant receptors expressed with the Semliki Forest virus system. *J Recept Signal Transduct Res* 19: 59–74
16 Gobbi M, Moia M, Pirona L, Morazzoni P, Mennini T (2001) *In vitro* binding studies with two *Hypericum perforatum* extracts – hyperforin, hypericin and biapigenin – on 5-HT_6, 5-HT_7, $GABA_A$/benzodiazepine, sigma, NPY-$Y_1/_2$ receptors and dopamine transporters. *Pharmacopsychiatry* 34 (Suppl 1): S45–S48
17 Butterweck V, Nahrstedt A, Evans J, Hufeisen S, Rauser L, Savage J, Popadak B, Ernsberger P, Roth BL (2002) *In vitro* receptor screening of pure constituents of St. John's Wort reveals novel interactions with a number of GPCRs. *Psychopharmacology* 162: 193–202
18 Klusa V, Germane S, Nöldner M, Chatterjee SS (2001) *Hypericum* extract and hyperforin: Memory-enhancing properties in rodents. *Pharmacopsychiatry* 34 (Suppl 1): S61–S69
19 Gambarana C, Tolu PL, Masi F, Rinaldi M, Giachetti D, Morazzoni P, De Montis MG (2001) A study of the antidepressant activity of *Hypericum perforatum* on animal models. *Pharmacopsychiatry* 34 (Suppl 1): S42–S44
20 Bhattacharya SK, Chakrabarti A, Chatterjee SS (1998) Activity profiles of two hyperforin-containing *Hypericum* extracts in behavioural models. *Pharmacopsychiatry* 31 (Suppl 1): 22–29
21 Müller WE, Singer A, Wonnemann M, Hafner U, Rolli M, Schäfer C (1998) Hyperforin represents the neurotransmitter reuptake inhibiting constituent of *Hypericum* extract. *Pharmacopsychiatry* 31 (Suppl 1): 16–21
22 Coleta M, Campos MG, Cotrim MD, Porenca da Cunha A (2001) Comparative evaluation of *Melissa officinalis* L., *Tilia europaea* L., *Passiflora edulis* Sims. and *Hypericum perforatum* L. in the elevated plus maze anxiety test. *Pharmacopsychiatry* 34 (Suppl 1): S20–S21
23 Kumar V, Jaiswal AK, Singh PN, Bhattacharya SK (2000) Anxiolytic activity of Indian *Hypericum perforatum* Linn: An experimental study. *Indian J Exp Biol* 38: 36–41
24 Vandenbogaerde A, Zanoli P, Puia G, Truzzi C, Kamuhabwa A, De Witte P, Merlevede W, Baraldi M (2000) Evidence that total extract of *Hypericum perforatum* affects exploratory behavior and exerts anxiolytic effects in rats. *Pharmacol Biochem Behav* 65: 627–633
25 Flausino OA Jr, Zangrossi H Jr, Salgado JV, Viana MB (2002) Effects of acute and chronic treatment with *Hypericum perforatum* L. (LI 160) on different anxiety-related responses in rats. *Pharmacol Biochem Behav* 71: 251–257
26 Beijamini V, Andreatini R (2003) Effects of *Hypericum perforatum* and paroxetine in the mouse defense test battery. *Pharmacol Biochem Behav* 74: 1015–1024
27 Beijamini V, Andreatini R (2003) Effects of *Hypericum perforatum* and paroxetine on rat performance in the elevated T-maze. *Pharmacol Res* 48: 199–207
28 Kumar V, Singh PN, Muruganandam AV, Bhattacharya SK (2000) Effect of Indian *Hypericum perforatum* Linn on animal models of cognitive dysfunction. *J Ethnopharmacol* 72: 119–128
29 Kumar V, Khanna VK, Seth PK, Singh PN, Bhattacharya SK (2002) Brain neurotransmitter receptor binding and nootropic studies on Indian *Hypericum perforatum* Linn. *Phytother Res* 16:

210–216
30 Khalifa A (2001) *Hypericum perforatum* as a nootropic drug: Enhancement of retrieval of memory of passive avoidance conditioning paradigm in mice. *J Ethnopharmacol* 76: 49–57
31 Widy-Tyszkiewicz E, Piechal A, Joniec I, Blecharz-Klin K (2002) Long term administration of *Hypericum perforatum* improves spatial learning and memory in water maze. *Biol Pharm Bull* 25: 1289–1294
32 Krishtal O, Lozovaya N, Fisunov A, Tsintsadze T, Pankratov Y, Kopanitsa M, Chatterjee SS (2001) Modulation of ion channels in rat neurons by the constituents of *Hypericum perforatum*. *Pharmacopsychiatry* 34 (Suppl 1): S74–S82
33 Keller JH, Karas M, Müller WE, Volmer DA, Eckert GP, Tawab MA, Blume HH, Dingermann T, Schubert-Zsilavecz M (2003) Determination of hyperforin in mouse brain by high-performance liquid chromatography/tandem mass spectrometry. *Anal Chem* 75: 6084–6088
34 Singer A, Wonnemann M, Müller WE (1999) Hyperforin, a major antidepressant constituent of St. John's Wort, inhibits serotonin uptake by elevating free intracellular Na^+. *J Pharmacol Exp Ther* 290: 1363–1368
35 Wonnemann M, Singer A, Müller WE (2000) Inhibition of synaptosomal uptake of ^{3}H-L-glutamate and ^{3}H-GABA by hyperforin, a major constituent of St. John's Wort: The role of amiloride sensitive sodium conductance pathways. *Neuropsychopharmacology* 23: 188–197
36 Buchholzer ML, Dvorak C, Chatterjee SS, Klein J (2002) Dual modulation of striatal acetylcholine release by hyperforin, a constituent of St. John's Wort. *J Pharmacol Exp Ther* 301: 714–719
37 Kiewert C, Buchholzer ML, Hartmann J, Chatterjee SS, Klein J (2004) Stimulation of hippocampal acetylcholine release by hyperforin, a constituent of St. John's Wort. *Neuroscience Lett* 364: 195–198
38 Collerton D (1986) Cholinergic function and intellectual decline in Alzheimer's disease. *Neuroscience* 19: 1–28
39 Gold PE (2003) Acetylcholine modulation of neural systems involved in learning and memory. *Neurobiol Learn Memory* 80: 194–201
40 Catania MA, Firenzuoli F, Crupi A, Mannucci C, Caputi AP, Calapi G (2003) *Hypericum perforatum* attenuates nicotine withdrawal signs in mice. *Psychopharmacol* 169: 186–189
41 Chatterjee SS, Erdelmeier C, Klessing K, Marme D, Schächtele C (1999) Stable hyperforin salts: Method of producing them and their uses in the treatment of Alzheimer's disease. Int. Pat. PCT Wo 99/41220, August 19
42 Forestl B, Steiner B, Müller WE (2003) Enhancement of proteolytic processing of β-amyloid precursor protein by hyperforin. *Biochem Pharmacol* 66: 2177–2184
43 Müller WE (2003) Current St. John's Wort research from mode of action to clinical efficacy. *Pharmacol Res* 47: 101–109

Edited by W. E. Müller

Pharmacokinetics and biopharmaceuticals

Mario Wurglics and Manfred Schubert-Zsilavecz

Institute of Pharmaceutical Chemistry, J.W. Goethe University Frankfurt/Main, Marie-Curie-Str. 9, D-60439 Frankfurt/Main, Germany

Introduction

Hypericum perforatum (St. John's Wort, SJW) is among the favourite herbal drugs, and is the only herbal alternative to classical synthetic antidepressants, in the therapy of mild-to-moderate depression. Several clinical studies have been conducted to verify the effectiveness of ethanolic or methanolic extracts [1–3]. Hyperforin (see Fig. 1) is the main source of pharmacological effects caused by the consumption of alcoholic extracts of St. John's Wort in the therapy of depression. It exerts many effects comparable to that of synthetic antidepressants like the uptake inhibition of 5-HT, norepinephrine, and dopamine. But in addition, the GABA and L-glutamate uptake is inhibited in the same manner [4]. These effects are probably based on an increase of the intracellular Na^+-concentration, due to an altering of sodium conductive pathways [5–10].

However, several studies indicate that flavone-derivatives, e.g., rutin and also the naphthodianthrones hypericin and pseudohypericin, take part in the antidepressant efficacy. Thus neither rutin alone nor SJW extracts with a reduced rutin content are therapeutically effective in the animal model, in contrast to SJW extracts containing usual levels of rutin [11]. For a long time the eye-catching red substance hypericin was thought to be responsible for the antidepressant activity of the extract, but in *in vitro* models no activity could

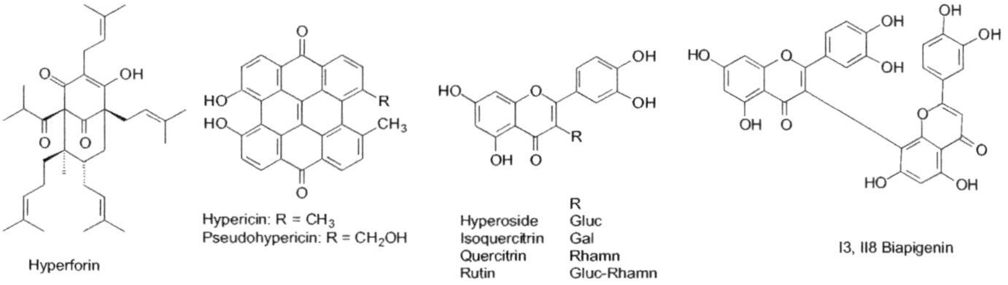

Figure 1. Chemical structure of ingredients from *Hypericum perforatum*.

be scientifically proved. Unlike the *in vitro* models, *in vivo* studies indicate that hypericin and pseudohypericin contribute to the extract's activity [12]. Apparently, the presence of cosolvents, like procyanidine B2 or hyperoside, plays a key role in hypericin's activity [13].

Even though the whole mechanism of action of alcoholic extracts of St. John's Wort has still not been clarified in detail, the antidepressant efficacy is well documented in various clinical studies.

Pharmacokinetics

In contrast to the good documentation concerning clinical efficacy, the oral bioavailability and the pharmacokinetic data about the active components are rather poor. Reasons for the lack of information are the absence of reference substances (e.g., ^{14}C-labeled molecules) and insufficient analytical methods. High performance liquid chromatography (HPLC) and UV methods are not sensitive enough for studies in humans. They can be used for the determination of active compounds in extracts or in some cases for animal studies. Proper pharmacokinetic studies in humans have been available since liquid chromatography/mass spectrometry/mass spectrometry (LC/MS/MS) methods were established.

Hyperforin

Hyperforin was presumed to be an unstable molecule and thus it was predicted that hyperforin could not be the active principle of *Hypericum perforatum* [14, 15]. However, Ostrowski published the first report concerning the oral bioavailability of hyperforin or hyperforin derivatives as early as 1988 [16]. The use of ^{14}C-labeled hyperforin, synthesized *in situ* in the plant and isolated by an elaboration procedure, made the qualitative detection of hyperforin in brain tissue possible. With the limitations of a HPLC/UV method, Biber et al. [17] were able to quantify hyperforin in the plasma of male Sprague Dawley rats after an oral dose of 300 mg/kg *Hypericum* extract WS 5572 (containing 5% hyperforin). This dose is well tolerated by the animals and reveals maximal antidepressant effects in behavioural models, but it is about 20-fold higher than the therapeutic dose in humans. Mean plasma C_{max} of hyperforin was reached 3 h after administration of the *Hypericum* extract and peaked at approximately 370 ng/ml. Further pharmacokinetic parameters could not be reliably estimated.

The implementation of HPLC-tandem mass spectrometry (TMS) after liquid–liquid extraction with hexane-ethyl acetate (70/30, v/v) allowed the detection of hyperforin in human plasma. The lower limit of quantification (1 ng/ml) made it possible to acquire pharmacokinetic parameters after a therapeutic dose of *Hypericum* extract in human. In a dose-ascending volunteer study, three doses of *Hypericum* extract WS 5572 (300, 600 and 1,200 mg) were evaluated. In a second repeated-dose study, 900 mg of extract WS 5572 were chosen as daily dose and compared to a daily dose of 900 mg extract WS 5573 (containing 0.5% hyperforin).

The results of the dose-ascending study demonstrated that after a lag time of approximately 1 h the absorption of hyperforin takes place (see Fig. 2). Nevertheless, the maximal plasma levels were attained within 3–3.5 h. The pharmacokinetic data is summarised in Table 1. A computer analysis of the data obtained resulted in an open two-compartment model with a half life of distribution and elimination half life of approximately 3 h and 9 h, respectively. No significant difference in the mean clearance could be found between the 300 and 600 mg dose. By contrast, the difference between 300 and 1,200 mg was statistically significant. This leads to the assumption that the pharmacokinetic of hyperforin is linear up to 600 mg doses of extract only.

In the repeated dose study, two extracts with different hyperforin content were compared for a period of 8 days. The data observed from the high hyperforin dose as well as from the low hyperforin dose study could be fitted to a two-compartment model. Independent of the hyperforin dose, the mean T_{max} values were comparable. However, the mean C_{max} and area under the curve (AUC) values after 900 and 1,200 mg doses were about 30% lower than those calculated from the three low dose studies, whereas the $t^1/_2$ values were of the same magnitude. The authors speculate that the loss of bioavailability is mainly due to the high lipophilicity of hyperforin or its interactions with some extract and/or gastrointestinal (GI) contents. In addition, the bioavailability of

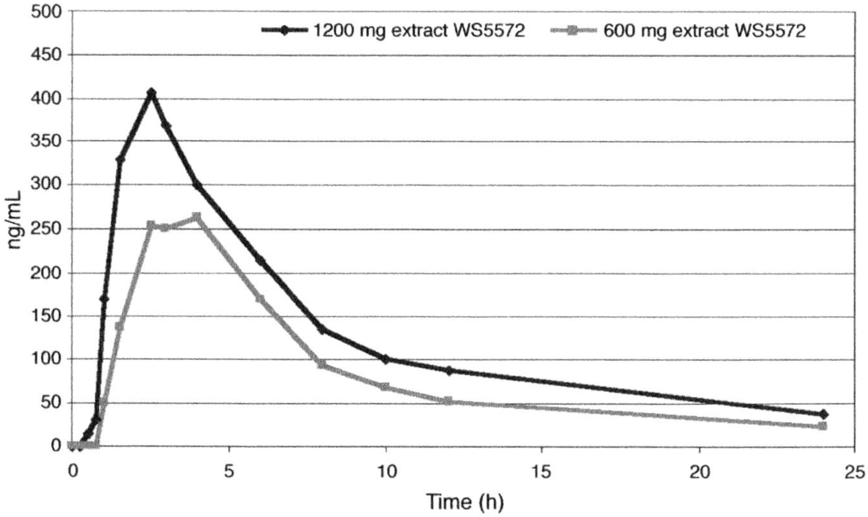

Figure 2. Plasma concentrations of hyperforin after oral administration of 600 and 1,200 mg *Hypericum* extract WS 5572 (5% hyperforin) in human [17].

Table 1. Pharmacokinetic parameters of hyperforin after single and multiple dose administration

Study	Dosage form	Extract	Extract/ dosage form [mg]	Hyperforin [mg]	C_{max} [ng/ml]	T_{max} [min]	AUC [ng·min/ml]	$t^1/_2$ [min]	CL [ml/min]
Biber [17]	Tablet	WS5572	300	14.8	153.15	214.8	1,335.9	567.6	199.3
Biber [17]	Tablet	WS5572	600	29.6	301.8	210	2,214.6	511.2	238.2
Biber [17]#	Tablet	WS5573	900	4.5	20.7	180	254	960	283
Biber [17]#	Tablet	WS5572	900	42.84	246	186	2336	672	266
Biber [17]	Tablet	WS5572	1,200	59.2	437.3	169.8	3,377.9	579	340.3
Franklin [18]	?	LI160	2,700	70*	1,450	210	-	-	-
Agrosi [20]	Soft gel capsule	-	300	15	168.35	150	1,482.7	-	-
Agrosi [20]	Hard gelatine capsule	-	300	15	84.25	184.8	583.65	-	-
Cui [21]	capsule	-	1,800	8.55	27.6	264		210	561.6

*Estimated according to reference [19].
Repeated dose study.

hyperforin decreased after repeated once-daily high doses of extract. On day 8, the mean C_{max} of hyperforin after the 900 mg WS 5572 dose was lower than on day 1. An increase in $t^{1/2}\,\beta$ observed by repeated doses could explain this loss of bioavailability.

In another high dose study, Franklin et al. [18] administered nine tablets of Jarsin® 300 (2,700 mg extract) to healthy volunteers and took plasma samples for the following 240 min. In contrast to the hyperforin content, the hypericin content of the tablets was previously determined (0.9 mg/tablet). The hyperforin content of the dose can only be estimated, based on later conducted batch-to-batch reproducibility study of this product [19]. The observed lag time of 1 h before the hyperforin absorption starts and the T_{max} values of this study were in good correlation to the data reported by Biber et al., while the C_{max} was slightly higher.

While in the aforementioned studies tablets were administered, Agrosi et al. compared the bioavailability of hyperforin and hypericin after a single dose of extract using hard gelatin (300 mg dry *Hypericum* extract per capsule) and softgel capsules (300 mg *Hypericum* extract in soy oil per capsule) [20]. The pharmacokinetic data obtained after administration of the softgel capsule was in good agreement with those reported by Biber et al., whereas C_{max} is only half as much in the volunteers treated with hard gelatin capsules. Cui et al. administered capsules with a low content of hyperforin [21]. The results were remarkably lower than that of Agrosi et al. Results of further pharmacokinetic studies are summarised in Table 1.

The plasma concentrations of approximately 150 ng/ml hyperforin measured in these studies are compatible with those Wonnemann et al. used in *in vitro* experiments to inhibit the synaptosomal uptake of serotonin, norepinephrine and dopamine [22]. However, these concentrations do not necessarily reflect those at the site of action e.g., the brain. Cervo et al. compared the antidepressant activity of two *Hypericum perforatum* extracts in relation to the hyperforin content (4.5% *versus* 0.5%) and the resulting plasma and brain concentrations in rats [23]. The effective dose (6.25 mg/kg i.p.) of the 4.5% hyperforin extract yielded in hyperforin plasma concentrations close to the plasma C_{max} found by Biber et al. in human volunteers after a daily dose of 900 mg extract (WS 5572). However, hyperforin concentrations in the brain did not achieve the lower limit of detection of the analytical method (about 20 ng/g). Sufficient hyperforin concentrations in brain tissue were obtained after at least three i.p. injections (within 24 h) of 12.5 mg/kg hyperforin dicyclohexylammonium salt. At this dose the mean whole brain concentration yielded 32.16 ± 16.08 ng/g (brain), while the plasma concentration peaked at 1.40 ± 0.24 μM. The calculated mean brain-to-plasma concentration ratio is about 0.04 ± 0.02.

Comparable data were obtained by Keller et al. after an oral dose of 15 mg/kg hyperforin sodium or an equivalent dose of *Hypericum* extract WS 5572 (containing 5% hyperforin) administered to mice [24]. For the first time, hyperforin was determined after an oral dose of a *Hypericum* extract in the brain of a rodent.

Naphthodianthrones – Hypericin, pseudohypericin

For a long time the naphthodianthrones (see Fig. 1) were thought to be the active principle of alcoholic *Hypericum* extracts in the therapy of mild-to-moderate severe depression. Some extracts are still standardised on the hypericin content. Early biochemical studies reported that hypericin is an inhibitor of monoamine oxidase-A and -B activity [25]. Later studies did not confirm these findings [26]. Nevertheless, hypericin showed efficacy in the Porsolt-test and some interesting endocrine effects [27].

Also well known is the phototoxicity, called hypericism, of hypericin. This is used in the photodynamic therapy (PDT), a promising new modality for the treatment of cancer. PDT involves the combination of a photosensitising agent (photosensitiser) and visible light of a wavelength matching the absorption spectrum of the drug. Hypericin is a powerful naturally occurring photosensitiser. Increased interest in hypericin as a potential clinical anticancer agent has arisen since several studies established its powerful *in vivo* and *in vitro* antineoplastic activity upon irradiation [28].

Additionally, antivirus activity against *Herpes simplex*, human immunodeficiency virus type 1 and some retroviruses has been described [29, 30].

It is therefore not surprising that the pharmacokinetic profile of hypericin and pseudohypericin has been thoroughly investigated. Since Liebes et al. determined hypericin in biological fluids in mice [31], several pharmacokinetic studies in humans have been carried out [18, 32–37].

Although hypericin and pseudohypericin (see Fig. 1) are structurally similar, their absorption rates after oral administration differ substantially (see Fig. 3). The plasma concentration of hypericin rose much later than those of pseudohypericin. The absorption of hypericin started after a lag time of approximately 2 h, whereas almost no lag time was observed for pseudohypericin (approximately 0.4 h). Hypericin levels peaked at an average of 5.8 h, while mean T_{max} of pseudohypericin was measured after 3.1 h (see Tab. 2). According to the higher levels of pseudohypericin in the extract (LI160), the maximum plasma levels of pseudohypericin were about twice those of hypericin in all studies, with one exception. Brockmöller et al. detected higher C_{max} values for hypericin, although the hypericin content in the dosage form was lower than the pseudohypericin content [35].

A calculation of the data resulted in a mean terminal elimination half life of about 32.7 h for hypericin and of 21.1 h for pseudohypericin. This was reflected by the fact that, even after 72 h, there were still measurable quantities of hypericin in plasma, while pseudohypericin had been completely eliminated by this time.

Steady-state pharmacokinetics were investigated in two studies. Kerb et al. [34] administered one Jarsin® 300 tablet t.i.d. whereas Brockmöller et al. [35] applied two Jarsin® 300 tablets t.i.d. The steady-state level was achieved after 6–7 days for hypericin and after about 4 days for pseudohypericin in both studies. For both substances the kinetic parameters – C_{max}, $t^1/_2$, clearance and

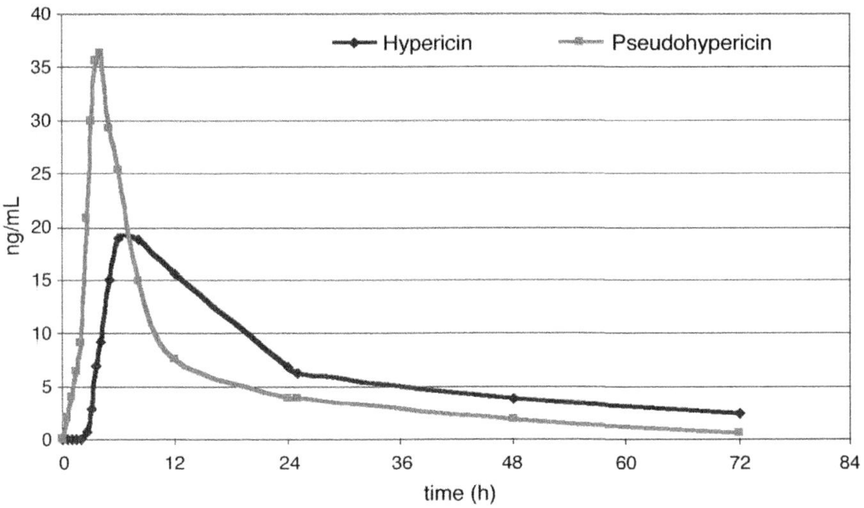

Figure 3. Plasma concentrations of hypericin and pseudohypericin after oral administration of 6 tablets of LI160 (1,800 mg *Hypericum* extract) in human [36].

AUC – were similar to those after a comparable single dose. The median through levels were, corresponding to $t^1/_2$ values, about two times higher for hypericin than for pseudohypericin, whereby distinct intraindividual variations were noted.

Intravenous administration of the naphthodianthrones permitted a rough estimation of the systemic availability, 14% for hypericin and 21% for pseudohypericin.

In summary, the plasma C_{max} of hypericin and pseudohypericin after administration of comparable amounts of *Hypericum* extracts was approximately 10-fold lower than maximum hyperforin plasma levels. In addition, the penetration of hypericin across the blood–brain barrier seems to be very limited. Fox et al. [38] investigated the cerebrospinal fluid (CSF) penetration in non-human primates after an i.v. bolus dose of at least 2 mg/kg (hypericin content of approximately 6–10 SJW dosage forms). There was no detectable hypericin in the CSF after an i.v. dose of 5 mg/kg, but phototoxic skin reaction could be observed.

A reason for the low bioavailability of hypericin and pseudohypericin is the poor water solubility of the naphthodianthrones. The presence of procyanidines or flavonol glycosides, present in *Hypericum* extracts, enhances the water solubility of both hypericins. In a pharmacokinetic study in rats, hyperoside increased the bioavailability of hypericin after oral administration by 34%. For procyanidin B2 a rise in bioavailability of hypericin of about 58% was detected [13].

Table 2. Pharmacokinetic parameters of hypericin and pseudohypericin after single and multiple dose administration

Study	Extract	Dosage form	Extract/ dosage form	Hypericin [mg]	C_{max} [ng/ml]	T_{max} [h]	AUC [ng·min/ml]	$t^{1}/_{2}$ [h]	CL [ml/min]	lag time [h]
Weiser [37]	LI160	Sugar coated tablet	300, 600, 1,200		4.3	2.5		6		
Staffeldt [36]	LI160	Sugar coated tablet	300	0.25	1.5	5.2	1,920	24.8		2.6
Staffeldt [36]	LI160	Sugar coated tablet	900	0.75	7.5	4.1	8,820	26		2
Staffeldt [36]	LI160	Sugar coated tablet	1,800	1.5	14.2	5.9	23,090	26.5		2.6
Brockmöller [35]	LI160	Sugar coated tablet	900	1.09	18	7.1	7,250	27.8	45	
Brockmöller [35]	LI160	Sugar coated tablet	1,800	2.18	36	6	59,580	29.1	40	
Brockmöller [35]	LI160	Sugar coated tablet	3,600	4.36	91	6.5	150,180	27.5	36.6	
Brockmöller [35]+	LI160	Sugar coated tablet	3 × 600	2.18#	29		16,200	41.7	44.7	
Kerb [34]	LI160	Sugar coated tablet	300	0.25	1.3	5.5	2,484	24.5	101	2.1
Kerb [34]	LI160	Sugar coated tablet	900	0.75	7.2	6	11,880	43.1	63.3	1.9
Kerb [34]	LI160	Sugar coated tablet	1,800	1.5	16.6	5.7	29,640	48.2	51	1.9
Kerb [34]+	LI160	Sugar coated tablet	3 × 300	0.75#	8.8		3,690	41.3	68.2	
Franklin [18]	LI160	Sugar coated tablet	2,700	8.1*						2.5
Agrosi [20]		capsule								
Bauer [32]	?	?	?	?	3.3	6				2
Pirker [33]	Nature's Way	capsule	?	?	?	?	?	?	?	?

* Total hypericin (hypericin + pseudohypericin); # Daily dose; + Repeated dose study.

(Continued on next page)

Table 2. (Continued)

Study	Extract	Dosage form	Extract/ dosage form	Pseudoypericin [mg]	C_{max} [ng/ml]	T_{max} [h]	AUC [ng·min/ml]	$t^{1}/_{2}$ [h]	CL [ml/min]	lag time [h]
Weiser [37]	LI160	Sugar coated tablet	300, 600, 1,200							
Staffeldt [36]	LI160	Sugar coated tablet	300	0.526	2.7	2.7	1,900	16.3		0.6
Staffeldt [36]	LI160	Sugar coated tablet	900	1.58	11.7	3	7,130	36		0.4
Staffeldt [36]	LI160	Sugar coated tablet	1,800	3.16	30.6	3.2	19,910	22.8		0.4
Brockmöller [35]	LI160	Sugar coated tablet	900	1.72	10	3.3	8,280	19.4	235	
Brockmöller [35]	LI160	Sugar coated tablet	1,800	3.44	25	3.2	20,040	16.1	205	
Brockmöller [35]	LI160	Sugar coated tablet	3,600	6.89	68	3.5	50,100	17.5	190	
Brockmöller [35]+	LI160	Sugar coated tablet	3 × 600	3.44#	29		9,900	22.8	116	
Kerb [34]	LI160	Sugar coated tablet	300	0.526	3.4	3	2,700	18.2	195	0.5
Kerb [34]	LI160	Sugar coated tablet	900	1.578	12.1	3	8,400	24.8	188	0.4
Kerb [34]	LI160	Sugar coated tablet	1,800	3.156	29.7	3	17,100	19.5	185	0.4
Kerb [34]+	LI160	Sugar coated tablet	3 × 300	1.578#	8.5		3,054	18.8	172	
Franklin [18]	LI160	Sugar coated tablet	2,700							
Agrosi [20]		capsule								
Bauer [32]	?	?	?	?	3.3	2				
Pirker [33]	Nature´s Way	capsule	?	?	?	?	?	?	?	?

Daily dose; + Repeated dose study.

Flavonoids

Flavonols are a large family of polyphenolic compounds that occur ubiquitously in the plant kingdom and thus are ingested by humans and animals with their regular diet. In the following section, we focus on the flavonoid sugar compounds, which are called glycosides, present in *Hypericum perforatum*: rutin, quercitrin, isoquercitrin and hyperoside (see Fig. 1). Quercetin, a representative of the flavonoid subclass of flavonols, is the common aglycone of these compounds. Also present in SJW extracts are the biflavones biapigenin and amentoflavone.

Some epidemiological studies point out positive effects of high intake of flavonoids on the prevention of cardiovascular diseases [39, 40]. The flavonoid glycosides are also considered to be efficacious in the treatment of chronic venous insufficiency [41, 42]. The protective effects observed seem to be a result of the antioxidative and radical scavenging capacity that quercetin showed in many *in vitro* experiments. However, clinical studies that support the use of quercetin or quercetin glycosides in the treatment of the above-mentioned conditions are few.

The role of the flavonol glycosides in the treatment of depression using St. John's Wort extracts is completely unclear. Nevertheless, SJW extracts without rutin showed less therapeutical efficacy than extracts containing usual amounts of rutin, while rutin alone did not exhibit efficacy in animal models [11].

Until now, there has been no evidence that hyperoside takes part in the antidepressant activity of SJW extracts, but hyperoside obviously improves the water solubility of the naphthodianthrone hypericin and increases its bioavailability by 34%. To what extent such interactions influence the bioavailability of active compounds of *Hypericum perforatum*, and furthermore the efficacy of the extract, has to be investigated in future experiments.

Up to the present, neither the pharmacokinetic profile nor the bioavailability of the flavonoids rutin, isoquercitrin, quercitrin, hyperoside and the biflavone biapigenin have been investigated after oral administration of a therapeutic dose of an alcoholic *Hypericum* extract in humans. Some pharmacokinetic data is available for rutin, administered as buckwheat tea or as pure rutin, for the aglycone quercetin and for other flavonol glycosides, administered in the form of onions (see Tab. 3) [43–48].

The mechanism of absorption is still the subject of lively discussion. Recent studies indicate that the sugar moiety attached to quercetin plays an important role. The absorption of the quercetin was significantly lower after the oral intake of the aglycone compared to the intake of various monoglucosides (e.g., quercetin-3-glucoside, quercetin-4'-glucoside). However, the bioavailability of quercetin after the ingestion of diglycosides (e.g., rutin) was about half that after the intake of monoglucosides. In addition, T_{max} from rutin was also significantly delayed (0.7 h for the monoglucosides *versus* 7 h for rutin), which indicates an absorption in the terminal ileum after microbial degradation. In

Table 3. Pharmacokinetic parameters of rutin and quercetin after single dose administration

Study	Source	Quercetin [mg]	C_{max} [µg/ml]	T_{max} [h]	AUC [µg·min/ml]	$t^{1/2}$ [h]	CL [l/h]
Graefe [43]	Onions	100	2.31	0.68	582	10.9	13.3
	Q-4'-O-glucosid	100	2.12	0.7	504	11.9	17.4
	Buckwheat tea	200	0.64	4.32	228	10.3	131
	Pure rutin	200	0.32	6.98	150	11.8	159
Erlund [45]	Pure quercetin	8	0.0414	1.9	37.8	17.1	
	Pure quercetin	20	0.0661	2.7	63.5	17.7	
	Pure quercetin	50	0.0861	4.9	79.4	15.1	
	Pure rutin	~8 (16 mg rutin)	0.0235	6.5	28.7	?	
	Pure rutin	~20 (40 mg rutin)	0.0476	7.4	47.58	?	
	Pure rutin	~50 (100 mg Rutin)	0.0899	7.5	72.12	?	
Erlund [44]	Pure quercetin	25	0.02–0.044				
Hollman [47]	Onions	68	0.224	0.70	139.8	28 ± 92	
	Äpple	98	92	2.51	62.8	23 ± 32	
	Pure rutin	100	90	9.3	59	-	
Ishii [50]	Pure rutin	500	0.063	5		2.5	

contrast, the uptake of quercetin glucosides within 1 h point at a resorption from the upper small intestine. Two hypotheses have been proposed for the explanation of this phenomenon: 1) Carrier mediated uptake of quercetin glucosides by the intestinal sodium-dependent glucose transporter SGLT1 [49]; 2) Deglucosilation by the enzyme phloridzin hydrolase in the brush border membrane and penetration of the lipophilic aglycone through the enterocyte membrane by passive diffusion [51].

Based on these data, it seems implausible that intact rutin reaches the plasma after a therapeutic dose of SJW extract (see Fig. 4). In these studies, rutin was not present in plasma; also free quercetin was not detectable, or detectable only in traces. Graefe et al. identified that, instead of intact rutin and free quercetin, four to five quercetin glucuronides were in human plasma. Furthermore, isorhamnetin (3'-*O*-methylquercetin) conjugates could be detected in human plasma. The total concentrations of isorhamnetin conjugates were about one-tenth of total quercetin concentrations [43]. After consumption of fried onions containing quercetin-3,4'-glucoside, quercetin-3-glucoside, quercetin-4'-glucoside and 3'-methylquercetin-4'-glucoside, Day et al. identified seven different quercetin and 3'-methylquercetin conjugates in human plasma, whereas the genuine compounds were not detectable [48].

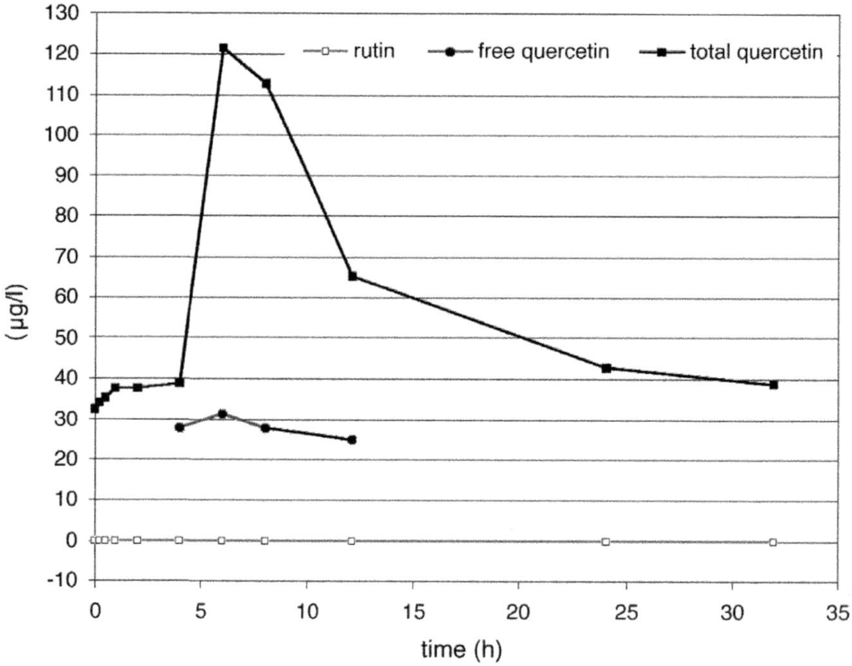

Figure 4. Plasma concentrations of total quercetin, unconjugated quercetin and rutin after administration of 100 mg rutin [45].

Major circulating compounds were quercetin-3-glucuronide, 3'-methyl-quercetin-3-glucuronide and quercetin-3'-sulfate. Overall, approximately 33% of the absorbed quercetin was present as sulfate conjugates and a fifth was methylated.

Free rutin attained measurable plasma levels only after an oral dose of 500 mg pure rutin [50]. Traces of tamarixetin (4'-*O*-methylquercetin) were also present in the plasma of rats [46].

A study in pigs demonstrated that all administered flavonols undergo an almost complete metabolism in the intestinal mucosa, because no free flavonols were detectable in portal plasma [51]. To what extent the intestinal microflora is involved in the degradation of the flavonol glycosides is still unclear. O´Leary et al. proposed the metabolism of quercetin-7-glucuronide/quercetin-3-glucuronide using a hepatic *in vitro* model. The metabolism of the glucuronides involved deglucuronidation by β-glucoronidase, methylation by the enzyme COMT, possible reglucuronidation by the enzyme UDP-glucuronosyltransferase and sulfation by sulfotransferase [52].

Conclusion

In conclusion, clinical studies performed using SJW extract with different concentrations of hyperforin have suggested that the antidepressive effect of St. John's Wort depends on its hyperforin content [2, 3]. The observed hyperforin plasma concentrations in humans reached about 300 ng/mL after oral administration of 600 mg extract WS 5572 [17]. This is very close to the concentrations of hyperforin that inhibit the synaptosomal uptake of serotonin *in vitro*. Furthermore, hyperforin is the only ingredient of *Hypericum perforatum* that could be determined in the brain of rodents after oral administration of alcoholic extracts.

The plasma concentrations of the hypericins were, compared to hyperforin, only a tenth, and until now, the hypericins were not found in the brain after oral administration of alcoholic *Hypericum* extracts or pure hypericin. Moreover, there is no evidence that the hypericins are able to pass the blood–brain barrier and reach the central nervous system (CNS). A recent study showed a better bioavailability of hypericin in presence of flavonoids and/or procyanidins, but further investigations are necessary to clarify the importance of these results.

The class of flavonoids is widespread in the plant kingdom. In *Hypericum perforatum,* the main compounds are the flavonol glycosides rutin, quercitrin, isoquercitrin and hyperoside. Until now, neither the pharmacokinetic profile nor the bioavailability of these flavonoids have been investigated after oral administration of a therapeutic dose of an alcoholic *Hypericum* extract in humans. Pharmacokinetic data is only available for rutin, administered as buckwheat tea or as pure rutin, for the aglycone quercetin and for other flavonol glycosides, administered in the form of onions. These studies indicate that flavonol glycosides like rutin are not absorpt intact after an oral dose, but appropriate concentrations of the aglycone quercetin can be detected.

In addition, the role of the flavonoids in the therapy of mild-to-moderate depression is still unclear. There is some evidence for the need of attendance of rutin in SJW-extracts, but the mechanism is not yet clarified.

Biopharmaceuticals

Introduction

In general, the biopharmaceutical quality and behaviour of herbal medicinal products (HMPs) is not well documented. The complexity of HMPs makes the *in vitro/in vivo* characterisation much more complicated than in chemically defined drug products. Only in a few cases, the active pharmaceutical ingredient is known – in many, if not most cases, it is suspected that several components may play a role. Further, the pharmacokinetics and bioavailability of HMPs are seldom described in literature.

Moreover, the regulatory situation in the developed countries with respect to quality control of HMPs various considerably. Although in a few countries (e.g., Germany) the quality requirements for dosage forms containing single active entities are also in general applicable to herbal medicinal products, in other countries (e.g., US) the main drug regulatory agency has little influence over HMPs, since many are introduced to the market as nutritional supplements.

By contrast, the European monograph (still under discussion) proposes three categories for herbal medicinal products according to how well their active components have been characterised [53]. For category A (standardised extracts; the active principle is identified) HMPs dissolution tests for the active components are required. If the active ingredient is known to be highly soluble in aqueous solutions at pH values typical of the GI tract, a disintegration test may be substituted for the dissolution test [54]. Further, the European Agency for the Evaluation of Medicinal Products (EMEA) does not require dissolution tests for HMPs falling under category B (quantified extracts; contain chemically defined constituents with relevant pharmacological properties (active markers)) or C (other extracts) if the extract is formulated as an immediate release product.

The EMEA concept should be extended for category B extracts, where active markers are known. This makes sense especially if the chosen marker is not very soluble in aqueous systems and therefore may serve as a true indicator of potential dissolution problems [55]. Guidelines for dissolution testing of oral dosage forms have been developed by the Food and Drug Administration (FDA) and the Fédération Internationale Pharmaceutique (FIP) [55–57].

In the case of St. John's Wort extract, three groups of components are though to be responsible for the extracts activity – the lipophilic phloroglucine derivative, hyperforin, the naphthodianthrones hypericin and pseudohypericin, which tend to exhibit medium polarity, and the more hydrophilic flavonoids and biflavones.

In order to determine how useful dissolution testing might be in characterizing the biopharmaceutical properties of St. John's Wort products, the dissolution of the active markers (hyperforin, hypericins, flavonoids and the biflavone biapigenin) was characterised to cover the entire spectrum of polarities present in the extract [58, 59]. In addition to standard compendial media, recently developed biorelevant media [60] were used in an attempt to better simulate *in vivo* release and therefore to better predict the influence of the formulation on the bioavailability of the extract.

Dissolution of hyperforin

Dissolution was conducted in Simulated Gastric Fluid USP 25 (without pepsin) (SGF_{sp}), since the stomach is the first part of the GI tract to come into contact with the dosage form. In this medium, no hyperforin dissolution was detected (see Fig. 5).

In conditions simulating the fasted state in the proximal small intestine, dissolution of hyperforin was also poor, about 5%. Studies comparing dissolution in FaSSIF in the presence and absence of the bile components (sodium taurocholate and lecithin) revealed that most of the (modest) increase in the % release compared to SGF_{sp} could be attributed to the bile components. When the concentration of bile components was increased further to approximate

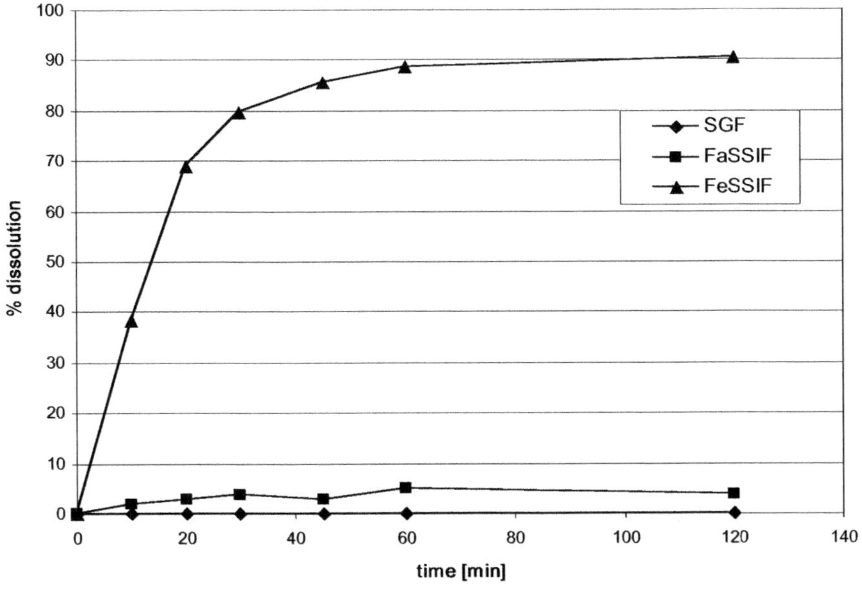

Figure 5. Dissolution of hyperforin from Texx 300® in different media.

postprandial conditions in the proximal small intestine, the dissolution of hyperforin was substantially improved, with 90% release from Texx® 300 within 2 h. The results for the other products are compared with those of Texx® 300 in Figure 6. The results indicate that even in the medium with the most advantageous composition for the release of hyperforin, some of the products performed poorly. Neither Jarsin® 300 nor Felis® 425 released more than 50% of their hyperforin content within 2 h of testing in FeSSIF.

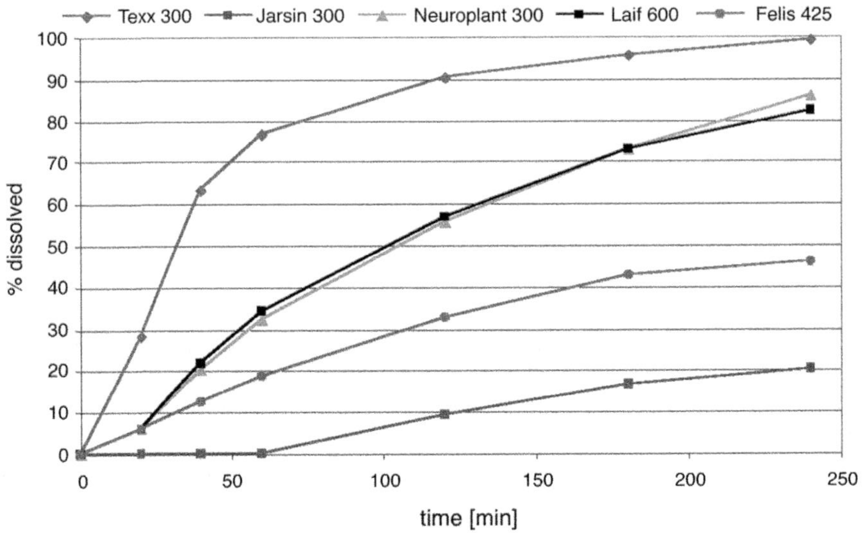

Figure 6. Dissolution of hyperforin in FeSSIF

Dissolution of total hypericin

As for hyperforin, no dissolution of hypericins could be detected in SGF. In FaSSIF the dissolution was considerably better, with up to 55% of the total hypericin content released within 2 h. This percentage may represent a falsely low value, due to an artefact of the experiment. Because five dosage form units per vessel had to be used to obtain quantifiable concentrations, some coning occurred at the bottom of the vessel. This phenomenon is known to inhibit dissolution and lead to poorer than expected results [61]. Due to analytical problems (interference of the medium), no results could be obtained in FeSSIF.

Dissolution of flavonoids

The dissolution of four flavonoids, rutin, hyperoside, isoquercitrin and quercitrin, were followed. Because the structures of hyperoside and isoquercitrin are closely related (see Fig. 1), complete separation by HPLC was not possible. Isoquercitrin concentrations were therefore calculated together with those for hyperoside and the results presented as the sum concentration. The relatively hydrophilic flavonoids dissolved well into all media tested.

To compare results among products, dissolution in FeSSIF was used, since this enabled us to simultaneously characterise the dissolution of both hyperforin and the flavonoids from the various products. Results for rutin are shown in Figure 7.

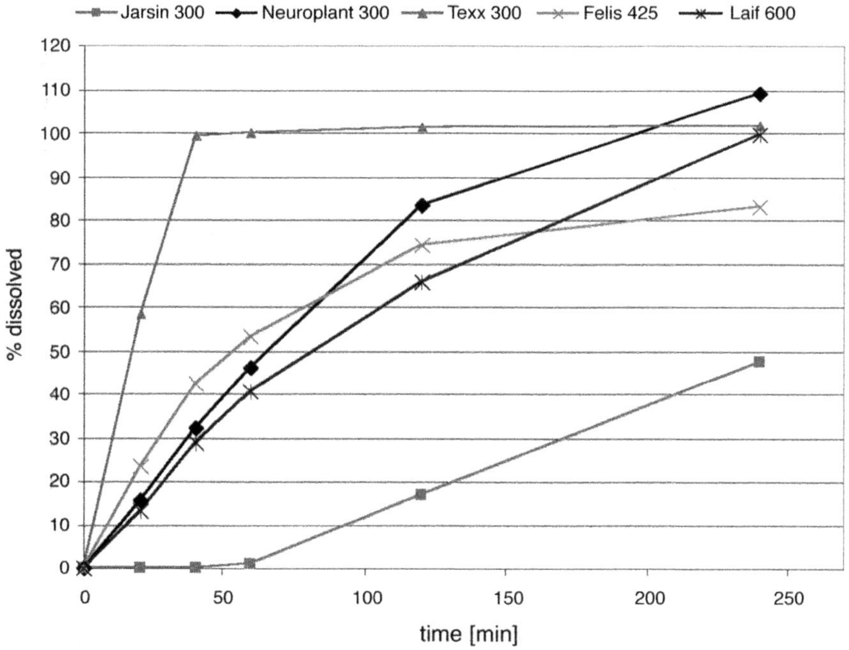

Figure 7. Dissolution of rutin in FeSSIF.

As for hyperforin, the products varied widely with respect to the release of rutin in FeSSIF. Whereas product Texx® 300 released 100% of its rutin content within 30 min, the product Jarsin® 300 released less than 50% within 4 h.

Dissolution of biapigenin

For biapigenin, best results were obtained using FeSSIF, representing the fed state conditions in the small intestine. 80% of biapigenin was released after 20 min of Laif® 900, compared to 70% using SGFsp. The reason that the dissolution profile of Laif® 900 using SGFsp has its maximum at 70% release is due to the fact that biapigenin is less soluble in SGFsp than in FaSSIF and

FeSSIF. The dissolution profile of the products using FaSSIF is not as uniform as using FeSSIF or SGFsp. However, biapigenin was released up to approximately 90% in FaSSIF after 90 min. Therefore the dissolution behaviour of all further St. John's Wort products were conducted in FeSSIF. The results are presented in Figure 8. After 120 min 100% biapigenin of Laif® 900, 90% of Felis® 650, 70% of Neuroplant® 300, 65% of Texx® 300 and 60% biapigenin of Felis® 425 was dissolved.

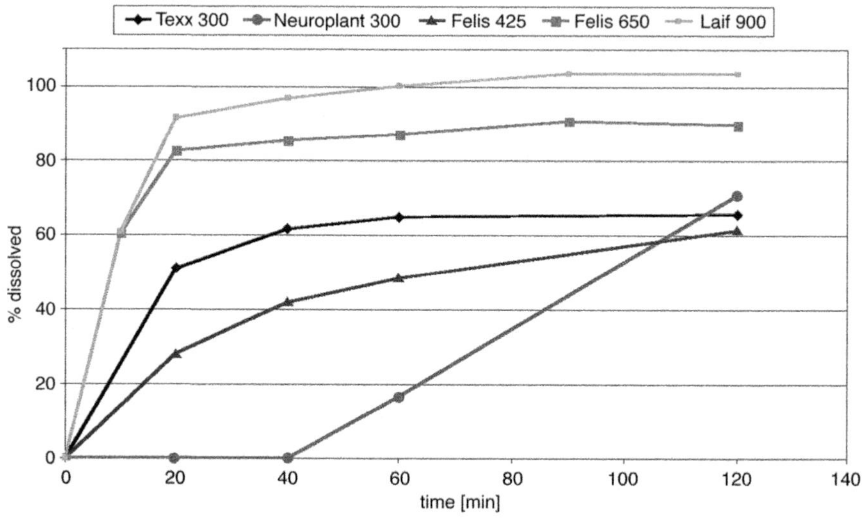

Figure 8. Dissolution of biapigenin in FeSSIF.

Discussion

The need for quality assurance, including confirmation of label strength, content uniformity and release properties, has long been recognised for drug products containing chemically defined, synthetically produced drugs. Relatively recently, a discussion about the extension of these concepts to HMPs has been initiated, with the result that it is now the norm (at least in Germany) for information such as the composition of the extraction fluid, the ratio of raw material to extraction fluid and the amount of extract in the product to appear on the label. Moreover, the better manufacturers strive to maintain batch-to-batch conformity of their herbal medicinal products by appropriate blending of extracts. Due partly to the variability in the raw material used to obtain the extract, however, it is not always possible to achieve the kind of content uniformity usually expected of products containing chemically defined actives [19]. Even when the batch-to-batch content conformity is good, this does not necessarily guarantee that the product will be bioequivalent, since absorption from the GI tract will depend at least partly on the release profile of the actives from the dosage form. Dissolution testing is the traditional yardstick of release from the dosage form, and in recent years the focus of dissolution testing has moved increasingly to the prediction of bioavailability and bioequivalence [62]. For these purposes, it is advantageous to simulate conditions in the GI tract in the *in vitro* test. Up until now, biorelevant dissolution testing has been used primarily for oral products containing single, chemically defined actives, but the time is ripe to consider applying these tests to the biopharmaceutical evaluation of HMPs belonging to categories A and B as well.

The advantages of the biorelevant media in comparison to typical compendial media for dissolution testing of herbal products are clearly demonstrated in this study. Although the dissolution of hydrophilic components such as the flavonoids can be adequately studied in simple aqueous media, the more lipophilic components (in the case of St. John's Wort, hypericins and hyperforin) are not sufficiently soluble in simple aqueous media to be able to compare release among products. The biorelevant media, which contain bile components at concentrations typically found in the GI tract, offer the possibility of comparing release among formulations in a way that should reflect the *in vivo* release. The dissolution results in the biorelevant media clearly demonstrate that there are glaring differences in the release properties of the various products studied and that these products could not be considered interchangeable. Since the dissolutions in FeSSIF were all conducted under sink conditions and the results were all normalised to the measured content of the various components in the products studied, the differences in the profile shapes must be attributed to differences in formulation among the products and not to differences in the product content.

According to the film model for dissolution [63], differences in the dissolution profiles could result from solubility, surface area, hydrodynamic or diffusivity differences among experiments. Of these factors, only the hydrodynamics held constant by standardising the test design in terms of volume of medium and stirring rate. Excipients in the formulation could influence solubility and wettability of the actives and perhaps exert a minor effect on the diffusivity of the actives. In addition, the particle size of the dry extracts used to manufacture the product could play a role in the release rate. Although the influence of formulation factors such as these on the biopharmaceutical properties of the product has been repeatedly documented for products containing chemically defined actives (especially those that are lipophilic and tend to exhibit poor dissolution), it is just beginning to be appreciated that these factors can be equally important for the *in vivo* performance of herbal medicinal products.

In the case of St. John's Wort, the evidence for the contribution of the lipophilic component, hyperforin, to the antidepressant activity is convincing. A comparison of the dissolution results in the various media indicates that the release of this component is far better under fed state (FeSSIF) conditions than under fasted state (SGF, FaSSIF) conditions. These results suggest that ingesting the dosage form with or after a meal may likely favour absorption of the hyperforin fraction of the extract.

Combined with our previous data concerning the batch-to-batch conformity of various St. John's Wort products [19], the dissolution data suggest that even when the labelling states that two products are pharmaceutically equivalent (e.g., Jarsin® 300 and Texx® 300), neither the hyperforin content nor its release from the dosage form under biorelevant conditions could be considered "essentially similar" for the two products. It is therefore recommended that patients taking St. John's Wort products should not switch from one product to another without considering the possibility that the dose may have to be adjusted to maintain the same therapeutic effect.

Conclusion

In this study, the advantages of the biorelevant media in comparison to typical compendial media for dissolution testing of herbal medicinal products are clearly demonstrated. The results of dissolution in biorelevant media show that there are glaring differences in the release properties of various products containing St. John's Wort extract. Therefore, it was concluded that these products could not be considered interchangeable. As a result, it is recommended that patients taking St. John's Wort products should not switch from one product to another without considering the possibility that the dose may have to be adjusted to maintain the same therapeutic effect.

References

1 Kasper S (2001) *Hypericum perforatum* – a review of clinical studies. *Pharmacopsychiatry* 34: 51–55
2 Berner M, Riemann D, Berger M (2002) The Efficacy of SJW extracts with more than 1% or less than 0.5% hyperforin content compared to placebo in the treatment of major depression. *Eur Arch Psy Clin Neur 252* (Suppl 1): 230
3 Laakmann G, Schüle C, Baghai T, Kieser M (1998) St. John`s Wort in mild to moderate depression: The relevance of hyperforin for the clinical efficacy. *Pharmacopsychiatry* 31 (Suppl 1): 54–55
4 Chatterjee SS, Bhattacharya SK, Wonnemann M, Singer A, Müller WE (1998) Hyperforin as a possible antidepressant component of *Hypericum* extracts. *Life Sci* 63: 499–510
5 Jensen AG, Hansen SH, Nielsen EO (2001) Adhyperforin as a contributor to the effect of *Hypericum perforatum* L. in biochemical models of antidepressant activity. *Life Sci* 68: 1593–1605
6 Müller WE, Singer A, Wonnemann M (2001) Hyperforin–antidepressant activity by a novel mechanism of action. *Pharmacopsychiatry* 34 (Suppl 1): S98–102
7 Singer A, Wonnemann M, Müller WE (1999) Hyperforin, a major antidepressant constituent of St. John's Wort, inhibits serotonin uptake by elevating free intracellular Na+1. *J Pharmacol Exp Ther* 290: 1363–1368
8 Wonnemann M, Singer A, Müller WE (2000) Inhibition of synaptosomal uptake of ^{3}H-L-glutamate and ^{3}H-GABA by hyperforin, a major constituent of St. John's Wort: the role of amiloride sensitive sodium conductive pathways. *Neuropsychopharmacology* 23: 188–197
9 Gobbi M, Moia M, Pirona L, Morizzoni P, Mennini T (2001) *In vitro* binding studies with two *Hypericum perforatum* extracts – hyperforin, hypericin and biapigenin – on 5-HT6, 5-HT7, GABA(A)/benzodiazepine, sigma, NPY-Y1/Y2 receptors and dopamine transporters.

Pharmacopsychiatry 34 (Suppl 1): S45–48
10 Gobbi M, Valle FD, Ciapparelli C, Diomede L, Morazzoni P, Verotta L, Caccia S, Cervo L, Mennini T (1999) *Hypericum perforatum* L. extract does not inhibit 5-HT transporter in rat brain cortex. *Naunyn Schmiedebergs Arch Pharmacol* 360: 262–269
11 Nöldner M, Schotz K (2002) Rutin is essential for the antidepressant activity of *Hypericum perforatum* extracts in the forced swimming test. *Planta Med* 68: 577–580
12 Butterweck V, Petereit F, Winterhoff H, Nahrstedt A (1998) Solubilized hypericin and pseudohypericin from *Hypericum perforatum* exert antidepressant activity in forced swimming test. *Planta Med* 64: 291–294
13 Butterweck V, Liefländer-Wulf U, Winterhoff H, Nahrstedt A (2003) Plasma levels of hypericin in presence of procyanidin B2 and hyperoside: A pharmacokinetic study in rats. *Planta Med* 69: 189–192
14 Hänsel R, Keller K, Rimpler H, Schneider G (Hrsg) (1993) Hager's Handbuch der Pharmazeutischen Praxis, Vol. 5. Springer Verlag, Berlin, Heidelberg, New York, 474–495
15 Hölzl J, Sattler S, Schütt H (1994) Johanniskraut: eine Alternative zu synthetischen Antidepressiva? *Pharm Ztg* 139: 3959–3977
16 Ostrowski E (1988) Untersuchung zur Analytik, ^{14}C-Markierung und Pharmakokinetik phenolischer Inhaltsstoffe von *Hypericum perforatum* L. Dissertation, Marburg, 118 ff
17 Biber A, Fischer H, Romer A, Chatterjee SS (1998) Oral bioavailability of *Hypericum* extracts in rats and humans. *Pharmacopsychiatry* 31 (Suppl 1): 36–43
18 Franklin M, Chi J, McGavin C, Hockney R, Reed A, Campling G, Whale RWR, Cowen PJ (1999) Neuroendocrine evidence for dopaminergic actions of *Hypericum* extract (LI 160) in healthy volunteers. *Biol Psychiatry* 46: 581–584
19 Wurglics M, Westerhoff K, Kaunzinger A, Wilke A, Baumeister A, Dressman J, Schubert-Zsilavecz M (2001) Batch-to-batch reproducibility of St. John's Wort preparations. *Pharmacopsychiatry* 34 (Suppl 1): 151–156
20 Agrosi M, Mischiatti S, Harrasser PC, Savio D (2000) Oral bioavailability of active principles from herbal products in humans. A study on *Hypericum perforatum* extracts using the soft gelatin capsule technology. *Phytomedicine* 7: 455–462
21 Cui Y, Gurley B, Ang CYW, Leakey J (2002) Determination of hyperforin in human plasma using solid-phase extraction and high-performance liquid chromatography with ultraviolet detection. *J Chromatogr B* 780: 129–135
22 Wonnemann M, Singer A, Siebert B, Müller WE (2001) Evaluation of synaptosomal uptake inhibition of most relevant constituents of St. John's Wort. *Pharmacopsychiatry* 34: 148–151
23 Cervo L, Rozio M, Ekalle-Soppo CB, Guiso G, Morazzoni P, Caccia S (2002) Role of hyperforin in the antidepressant-like activity of *Hypericum perforatum* extracts. *Psychopharmacology* 164: 423–428
24 Keller JH, Karas M, Müller WE, Volmer DA, Eckert GP, Abdel Tawab M, Blume HH, Dingermann T, Schubert-Zsilavecz M (2003) Determination of hyperforin in mouse brain by high-performance liquid chromatography/tandem mass spectrometry. *Anal Chem* 75: 6084–6088
25 Suzuki O, Katsumata Y, Oya M, Bladt S, Wagner H (1984) Inhibition of monoamine oxidase by hypericin. *Planta Med* 3: 272–274
26 Demisch L, Hölzl J, Gollnik B, Kaczmarczyk P (1989) Identification of selective MAO-type A inhibition of *Hypericum perforatum* L. *Pharmacopsychiatry* 22: 194
27 Butterweck V, Korte B, Winterhoff H (2001) Pharmacological and endocrine effects of *Hypericum perforatum* and hypericin after repeated treatment. *Pharmacopsychiatry* 34 (Suppl 1): 2–7
28 Agostinis P, Vantieghem A, Merlevede W, de Witte PA (2002) Hypericin in cancer treatment: more light on the way. *Int J Biochem Cell Biol* 34: 221–241
29 Lenard J, Rabson A, Vanderoef R (1993) Photodynamic inactivation of infectivity of human immunodeficiency virus and other enveloped viruses using hypericin and rose bengal: inhibition of fusion and syncytia formation. *Proc Natl Acad Sci USA* 90: 158–162
30 Andersen DO, Weber ND, Wood SG, Hughes BG, Murray BK, North JA (1991) *In vitro* virucidal activity of selected anthraquinones and anthraquinone derivatives. *Antivir Res* 16: 185–196
31 Liebes I, Mazur Y, Freeman D, Lavie D, Lavie B, Kudler N, Mendoza S, Levin B, Hochster H, Meruelo D (1991) A method for the quantification of hypericin, an antiviral agent, in biological fluids by high-performance liquid chromatography. *Anal Biochem* 195: 77–85
32 Bauer S, Strömer E, Graubaum HJ, Roots I (2001) Determination of hyperforin, hypericin, and pseudohypericin in human plasma using high-performance liquid chromatography analysis with

fluorescence and ultraviolet detection. *J Chromatogr B* 765: 29–35

33 Pirker R, Huck CW, Bonn GK (2002) Simultaneous determination of hypericin and hyperforin in human plasma and serum using liquid-liquid extraction, high-performance liquid chromatography and liquid chromatography-tandem mass spectrometry. *J Chromatogr B* 777: 147–153

34 Kerb R, Brockmöller J, Staffeldt B, Ploch M, Roots I (1996) Single-dose and steady-state pharmacokinetics of hypericin and pseudohypericin. *Antimicrob Agents Chemother* 40: 2087–2093

35 Brockmöller J, Reum T, Bauer S, Kerb R, Hübner WD, Roots I (1997) Hypericin and pseudohypericin: pharmacokinetics and effects on photosensitivity in humans. *Pharmacopsychiatry* 30: 94–101

36 Staffeldt B, Kerb B, Brockmöller J, Ploch M, Roots I (1994) Pharmacokinetics of hypericin and pseudohypericin after oral intake of the *Hypericum perforatum* extract LI 160 in healthy volunteers. *J Geriatr Psychiatry Neurol* 7 (Suppl 1): 47–53

37 Weiser D (1991) Pharmakokinetic von Hypericin nach oraler Einnahme des Johanniskraut-Extraktes LI160. *Nervenheilkunde* 10: 318–319

38 Fox E, Murphy RF, McCully L, Adamson PC (2001) Plasma pharmacokinetics and cerebrospinal fluid penetration of hypericin in nonhuman primates. *Cancer Chemother Pharmacol* 47: 41–47

39 Hertog MG, Feskens EJ, Kromhout D (1997) Antioxidant flavonols and coronary heart disease risk. *Lancet* 349: 699

40 Yochum L, Kushi L, Meyer K, Folsom A (1999) Dietary flavonoid intake and risk of cardiovascular disease in postmenopausal women. *Am J Epidemiol* 149: 943–949

41 Kiesewetter H, Koscielny J, Kalus U, Vix JM, Peil H, Petrini O, van Toor BS, de Mey C (2000) Efficacy of orally administered extract of red vine leaf AS 195 (*Folia vitis viniferae*) in chronic venous insufficiency (stages I–II): a randomized, double-blind, placebo-controlled trial. *Arzneimittelforschung* 50: 109–117

42 Ihme N, Kiesewetter H, Jung F, Hoffmann KH, Birk A, Muller A, Grutzner KI (1997) Leg oedema protection from buckwheat herb tea in patients with chronic venous insufficiency: a single-centre, randomised, double-blind, placebo-controlled clinical trial. *Eur J Clin Pharmacol* 32: 427–436

43 Graefe EU, Wittig J, Mueller S, Riehtling AK, Uehleke B, Drewelow B, Pforte H, Jacobash G, Derendorf H, Veit M (2001) Pharmacokinetics and bioavailability of quercetin glycosides in humans. *J Clin Pharmacol* 41: 492–499

44 Erlund I, Alfthan G, Siren H, Arniemi K, Aro A (1999) Validated method for the quantitation of quercetin from human plasma using high-performance liquid chromatography with electrochemical detection. *J Chromatogr B* 727: 179–189

45 Erlund I, Kosonen T, Alfthan G, Mäenpää J, Perttunen K, Kenraali J, Parantainen J, Aro A (2000) Pharmacokinetics of quercetin from quercetin aglycone and rutin in healthy volunteers. *Eur J Clin Pharmacol* 56: 545–553

46 Manach C, Morand C, Demigne C, Texier O, Regerat F, Remesy C (1997) Bioavailability of rutin and quercetin in rats. *FEBS Lett* 409: 12–16

47 Hollman PCH, van Trijp JMP, Buysman NCP, v.d. Gaag MS, Mengelers MJB, de Vries JHM, Katan MB (1997) Relative bioavailability of the antioxidant flavonoid quercetin from various foods in man. *FEBS Lett* 418: 152–156

48 Day A, Mellon F, Barron D, Sarrazin G, Morgan MRA, Williamson G (2001) Human metabolism of dietary flavonoids: identification of plasma metabolites of quercetin. *Free radic res* 35: 941–952

49 Wolffram S, Blöck M, Ader P (2002) Quercetin-3-glucoside is transported by the glucose carrier SGLT1 across the brush border membrane of rat small intestine. *J Nutr* 132: 630–635

50 Ishii K, Furuta T, Kasuya Y (2001) Determination of rutin in human plasma by high-performance liquid chromatography utilizing solid-phase extraction and ultraviolet detection. *J Chromatogr B* 759: 161–168

51 Cermak R, Landgraf S, Wolffram S (2003) The bioavailability of quercetin in pigs depends on the glycoside moiety and on dietary factors. *J Nutr* 133: 2802–2807

52 O´Leary KA, Day AJ, Needs PW, Mellon FA, O´Brien NM, Williamson G (2003) Metabolism of quercetin-7- and quercetin-3-glucuronides by an *in vitro* hepatic model: the role of human beta-glucuronidase, sulfotransferase, catechol-O-methyltransferase and multi-resistant protein 2 (MRP2) in flavonoid metabolism. *Biochem Pharmacol* 65: 479–491

53 Lang F, Stumpf H. (1999) Considerations on future pharmacopoeial monographs for plant extracts. *Pharmeuropa* 11: 268–275

54 EMEA (1999) Note for guidance on specifications: Test procedures and acceptance criteria for herbal drugs, herbal drug preparations and herbal medicinal products. *http://www.eudra.org/emea.html* (Accessed June 2003)
55 Lang F, Keller K, Ihrig M, van Oudtshoorn-Eckard J, Möller H, Srinivasan S, He-ci Y (2003) Biopharmaceutical characterisation of herbal medicinal products. *Pharm Forum* 29: 1337–1346
56 Siewert M (1997) FIP guidelines for dissolution testing of solid oral products. *Pharm Ind* 59: 760–766
57 FDA Guidance for Industry (1997) Dissolution of immediate release solid oral dosage Forms. *http://www.fda.gov/cder/guidance.html* (Accessed June 2003)
58 Westerhoff K, Kaunzinger A, Wurglics M, Dressman J, Schubert-Zsilavecz M (2002) Biorelevant dissolution testing of St John's Wort products. *J Pharm Pharmacol* 54: 1615–1621
59 Schulte-Löbbert S, Westerhoff K, Wilke A, Schubert-Zsilavecz M, Wurglics M (2003) Development of a high-performance-liquid-chromatographic method for the determination of biapigenin in biorelevant media. *J Pharm Biomed Anal* 33: 53–60
60 Galia E, Nicolaides E, Hörter D, Löbenberg R, Reppas C, Dressman JB (1998) Evaluation of various dissolution media for predicting *in vivo* performance of class I and II drugs. *Pharm Res* 15: 698–705
61 Beckert A, Quach T (1996) Improved hydrodynamics for apparatus 2. *Dissolution Technologies* 3
62 Shah V (2001) Dissolution: A quality control test *versus* a bioequivalence test. *Dissolution Technologies* 8: 6–7
63 Dressman JB, Amidon GL, Reppas C, Shah VP (1998) Dissolution testing as a prognostic tool for oral drug absorption: immediate release dosage forms. *Pharm Res* 15: 11–22

St. John's Wort and its Active Principles in Depression and Anxiety
Edited by W. E. Müller

Clinical efficacy in depression

Siegfried Kasper

Klinische Abteilung für Allgemeine Psychiatrie, Universitätsklinik für Psychiatrie, Medizinische Universität Wien, Währinger Gürtel 18-20, 1090 Wien, Austria

Introduction

Depression is among the most frequent psychiatric disorders and is estimated to have a lifetime prevalence rate of 17% [1]. Although considerable research has been performed in this disorder to uncover the underlying biology and subsequent treatment, there is still an increase of the disease and the World Health Organization (WHO) estimates that by the year 2020 it will parallel the morbidity and economic loss caused by heart diseases [2].

Among the different pharmacotherapeutic approaches, St. John's Wort (SJW) established its role for the treatment of mild-to-moderate depression. This assumption was not supported 20 years ago but there has been a considerable number of studies from 1990 onwards in agreement with this conclusion. Treatment with the older antidepressants of the tri- and tetracyclic categories was characterised by an unacceptable amount of side effects and also toxicity in overdose. It was the merit of the modern antidepressants, spearheaded by the selective serotonin reuptake inhibitors (SSRIs), to exert antidepressant efficacy with a low side effect profile. Subsequently, randomised controlled trials exerted that the favourable side effect profile of the newer antidepressants is even better with preparations of SJW.

The aim of this review is to summarise the antidepressant effectiveness of SJW in the treatment of depression and to report on the tolerability and safety profile.

Short term studies

Efficacy

As outlined in Tables 1 and 2, there are a considerable number of placebo-controlled studies using different herbal extracts of SJW and most of them indicate its clinical efficacy. Whereas the earlier studies did not fulfil the methodological standards that are now established and required according to the "Good Clinical Practice" and acknowledged by the American (FDA) and

Table 1. Placebo-controlled clinical studies with LI 160 in depressive disorders since 1994

Author, year [Ref]	N Groups	Duration of treatment (days)	Response (HAMD)
Sommer & Harrer, 1994 [26]	25 LI 160 25 Placebo	28	67% diff. +39% 28%
Hübner et al., 1994 [27]	20 LI 160 19 Placebo	28	70% diff. +23% 47%
Hänsgen et al., 1994 [28]	51 LI 160 50 Placebo	28	70% diff. +46% 24%
Bjerkenstedt, 2000 [29]	54 LI 160 54 Fluoxetin 55 Placebo	28 (42)	33% diff. +13% 33% diff. +13% 20%
Montgomery et al., 2000 [30]	115 LI 160 120 Placebo	42 (84)	48% diff. ±0% 48%
Shelton et al., 2001 [15]	98 LI 160 102 Placebo	56	27% diff. +8% 19%
Davidson et al., 2002 [16]	112 LI 160 112 Placebo 112 Sertralin	56 (182)	43% diff. –7% 53% diff. +3% 50%

diff.: Difference of St. John's Wort preparations to placebo.

European (EMEA) health authorities, studies in the last 10 years, however, have been conducted according to these guidelines and compared the effects of SJW with placebo and/or synthetic antidepressants like imipramine, maprotiline and also with the newer antidepressants like fluoxetine and sertraline.

A number of reviews concluded that SJW extracts exert antidepressant efficacy in the treatment of depression [3–11]. In the Cochrane Report [7], 45 randomised studies were identified that judged the antidepressant efficacy of SJW; 18 of them could not be used based on methodological consideration, of the 28 remaining studies (including 2,291 patients) 17 were placebo-controlled (1,168 patients) and 10 (1,123 patients) compared SJW preparations with synthetic antidepressants. In the placebo-controlled studies, there was a response rate (50% reduction mostly used as a response criterion) of SJW preparations of 56% and 25% for placebo. Comparable results have been found for the response rates of antidepressants (52%) and SJW preparations (50%).

In a new Cochrane analysis [8], 27 trials were included with a total of 2,291 patients. Among them, 17 trials were placebo-controlled and 10 with active comparator compounds. These studies revealed that *Hypericum* preparations were significantly superior to placebo (rate ratio 2.47; 95% confidence interval 1.69 to 3.61) and similarly effective as standard antidepressants. However, the rate of side effects was lower for *Hypericum* single preparations (23.3%) compared to 44.7% for standard antidepressants. The authors concluded that

Table 2. Controlled clinical studies with SJW extracts other than LI 160 in depressive disorders since 1994

Author, year	N groups	Duration of treatment (days)	Extract	Response (HAMD)
Schrader et al., 1998 [31]	80 *Hypericum* 79 Placebo	42	ZE 117	56% diff. +41% 15%
Laakmann et al., 1998 [23]	49 Hyperforin 5% 49 Hyperforin 0.5% 49 Placebo	42	WS 5572 WS 5573	49% diff. +16% 39% diff. +6% 33%
Philipp et al., 1999 [32]	100 *Hypericum* 105 Imipramine 46 Placebo	56	STEI 300	76% diff. +13% 67% diff. +6% 63%
Harrer et al., 1999 [33]	70 *Hypericum* 79 Fluoxetine	42	LoHyp-57	71% diff. –1% 72%
Schrader, 2000 [34]	125 *Hypericum* 113 Fluoxetine	42	ZE 117	60% diff. +20% 40%
Woelk, 2000 [35]	157 *Hypericum* 167 Imipramine	42	ZE 117	43% diff. +3% 40%
Volz & Laux, 2000 [36]	70 *Hypericum* 70 Placebo	42	D-0496	67% diff. +15% 52%
Kalb et al., 2001 [12]	37 *Hypericum* 35 Placebo	42	WS 5572	23% diff. +8% 15%
Lecrubier et al., 2002 [13]	186 *Hypericum* 188 Placebo	42	WS 5570	53% diff. +10% 43%

diff: Difference to placebo or reference antidepressant.

Hypericum extracts are more effective than placebo for the short-term treatment of mild-to-moderately severe depressive disorders, however note that the current evidence is inadequate to establish whether *Hypericum* is as effective as other antidepressants. The authors further indicate the necessity to establish the antidepressant properties over longer observation periods.

From a "mode of action" point of view, an interesting series of studies has recently been carried out with SJW extracts containing a high content of hyperforin. Laakmann and colleagues (1998) demonstrated that a patient group receiving higher amounts of hyperforin exerted a significantly better response than the placebo group, a finding that was not obtained in the patient group receiving a low amount of hyperforin. This finding was replicated in the study of Kalb et al. (2001) [12] and Lecrubier et al. (2002) [13]. Whereas the studies of Laakmann et al. and Kalb et al. were carried out in Germany, the large study of Lecrubier (including 168 patients with SJW and 188 patients with placebo) was conducted in France. All three individual studies and the meta-analysis of these studies [14] revealed a positive response of SJW.

However, two recent studies [15, 16] could not demonstrate superiority of *Hypericum* extract over placebo, and in the study of Davidson et al. also compared to sertralin. The study of Shelton et al. included 200 tertiary care outpatients suffering from a single or recurrent episode of major depression according to DSM-IV criteria. Shelton et al. included chronic patients who had an average duration of depression of more than 2 years and who were treated as outpatients in tertiary care clinics in academic medical centres. Therefore, the population of the study of Shelton et al. differed substantially from the trials carried out in Europe in primary care settings with less chronically ill patients. This patient group also exhibited an unusually low responder rate of 20.7%, which speaks for the chronicity of the sample. Contrastingly, the placebo response rate in antidepressant trials is nowadays between 30% and 50%.

The study of Davidson et al. [16] compared in a randomised control design SJW (900 mg) with placebo and sertralin (50–100 mg) and could not demonstrate that either sertralin or SJW were better than placebo in the primary efficacy criteria. However, in a secondary measurement there was an advantage of sertralin over placebo, a finding not demonstrated for the SJW group. When the patients and doctors were asked about the possible content of the medication in the double-blind study they could, however, correctly identify sertralin, based presumably on the side effect profile. Therefore, the double-blind conditions were questionable in this study.

A number of drug observational studies confirmed the effectiveness and tolerability of SJW in a non-selected population of mild-to-moderate depression (e.g., Rychlik et al. [17]).

Tolerability and safety

One of the characteristics of SJW preparations is that they show a favourable side effect profile. This has been confirmed in all large meta-analyses and also in drug observational trials. Symptoms like nausea appeared with a frequency of 1.2% in the large drug observational study of Vorbach et al. [18] and the overall number of unwanted side effects was 3.01%. Although the side effects profile is lower in drug observational studies, considerably higher numbers are obtained with synthetic antidepressants in these designs.

There are no changes in electrocardiogram (ECG) parameters and no fatal intoxication has been reported with SJW extract monotherapy in the literature.

A few reports have been published about interactions of SJW preparations that indicate that SJW preparations can lower the plasma level of concomitantly used antidepressants which has been shown for amitriptyline and nortriptyline [19]. Furthermore, a possible interaction has been reported between SJW preparations and anticoagulants of the coumarin type. This resulted in a lower dosage of phenprocoumon [9]. The pathophysiological mechanism of this phenomenon is an induction of cytochrome P-450 isoenzymes and/or a blockade of intestinal resorption of phenprocoumon preparations. Furthermore, a lower concentration of digoxin has been reported in a placebo controlled study in which SJW was included [20]. Also a few cases of SJW induced menstrual bleeding were reported. A contraindication of the co-medication of SJW with cyclosporine has been confirmed [19], since there is a lowering of cyclosporine plasma levels [21]. A similar finding has been reported for the concomitant use of indinavir [22]. Based on this interaction profile, it is apparent that patients receiving SJW preparations and the above mentioned medications should be monitored carefully.

Meta-analysis

Meta-analytic approaches are one way to further elucidate the role of efficacy in treatment studies. Since SJW extract studies have been challenged for questionable methodology with regard to (1) failure to employ standardised diagnostic criteria, (2) failure to use a standardised symptom rating instrument, (3) a relatively short period of observation and (4) study conduct by insufficiently trained and experienced investigators, three studies have been selected that do not meet this criticism. In the study of Laakmann et al. [23], Kalb et al. [12] and Lecrubier et al. [13], all patients suffered from mild-to-moderate major depression according to DSM-IV criteria and the primary outcome measurement for treatment was the change in total score of the Hamilton Rating Scale for depression. Furthermore, a double-blind treatment period of 6 weeks was obtained in a placebo-controlled fashion and the raters who established the primary outcome measurement (HAM-D) underwent a rater training that covered also the diagnostic process.

The meta-analysis was performed on the original data set of the three double-blind, randomised placebo-controlled, multicentre trials including 554 outpatients (see Tab. 3). The relationship between the symptoms of depression as represented by the individual items of the HAM-D was assessed in a cluster analytic approach taking into account the individual items. This meta-analysis has been published recently by Kasper and Dienel [14] and revealed

Table 3. Characteristics of meta-analysis of three recent European studies [14]

- Same extract (WS 5572, WS 5570)
- Same methodology
  - *Standardized diagnosis (DSM-IV)*
  - *Standardized rating instruments (HAM-D)*
  - *Experienced psychiatrists*
  - *Primary care setting (No tertiary sample)*
  - *Mild to moderate depression (No severe sample!)*
  - *No chronic sample*
- On whole data set of individual patients
  - Laakmann et al., 1998 [23]
  - Lecrubier et al., 2002 [13]
  - Kalb et al., 2001 [12]

two clusters of items that were stable in several independent subsets of the full data set (see Fig. 1). Cluster 1 included the HAM-D items 1,2,3,7,8,12,13,14,16 and cluster 2 the items 4,5,6,9,10,11,15,17. Whereas cluster 1 was interpreted to represent the core symptoms of depression, cluster 2 was primarily composed of items assessing depression related anxiety and insomnia. It was apparent that in both clusters *Hypericum* extract reduced the symptoms of depression more effectively than placebo. The results of this meta-analysis indicate that the *Hypericum* extract used in this studies (WS® 5570 or WS® 5572), which was a hyperforin rich extract, treats the symptomatology of depression in a very specific way by influencing the core signs and symptoms of the disease comparable to modern synthetic antidepressants.

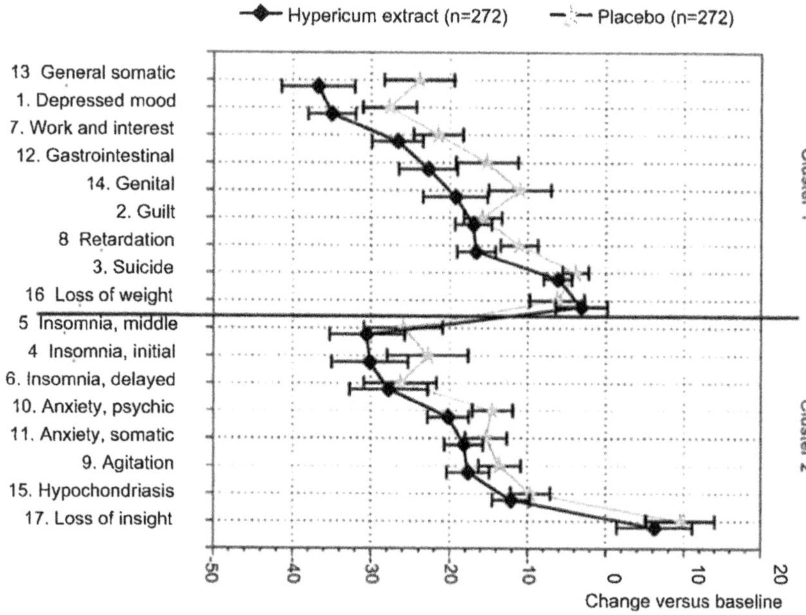

Figure 1. Clusters of items obtained with meta-analysis [14]. Rescaled item score change *versus* baseline. Negative values indicate symptom improvement; items ordered within clusters by descending magnitude of change in the *Hypericum* group); from: [14].

Long-term studies

Considerable research has been conducted in the long-term treatment of synthetic antidepressants since the European (EMEA), but not the North American (FDA), health authorities asked for long-term studies before licensing medication.

Until now there have been no long-term studies published in which SJW preparations were compared to placebo. First results from open long-term studies over the time span of 12 months [24] demonstrated that the antidepressant efficacy of SJW achieved in acute treatment was sustained over the subsequent year.

In an ongoing trial, patients with mild-to-moderate recurrent major depression are studied in a placebo-controlled design with regard to the prophylactic efficacy and safety of *Hypericum* extract WS®5570 which contains a standardised content of 3–6% hyperforin. The study is conducted in psychiatric as well as general primary care practices in Germany and Sweden and, to ensure uniformed diagnostic processes and ratings of depression, rater trainings were performed. The methodology of this study is outlined in the publication of Kasper et al. [25] and the design from Figure 2 indicates that patients receive a single blind acute treatment with 3 × 300 mg/day of WS®5570 for 6 weeks, those who respond to this acute treatment are then randomised at a ratio of 2:1 to 26 weeks of double-blind continuation treatment with the same dosage regimen. After completion of the continuation treatment, those patients who did not show a relapse start additional 52 weeks of double blind maintenance treatment. Based on this design, it is possible to have results on relapse (first 6 months) and recurrence (1 year) under treatment of a SJW preparation and placebo. Since long-term studies take a long time, the results will only be available in the near future.

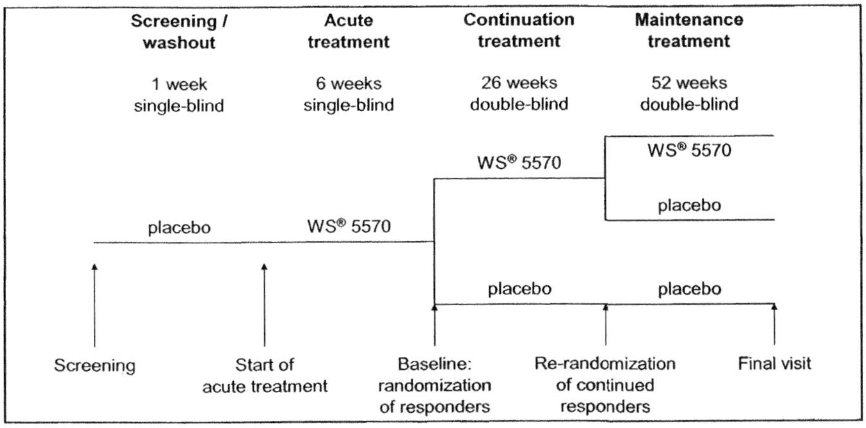

Figure 2. Study flow chart of placebo-controlled SJW long-term study [25].

Conclusion

Whereas 20 years ago SJW preparations were not taken seriously for the treatment of depression it is apparent that the available studies now, with a few exceptions, indicate that its use for indication of mild-to-moderate depression is justified. However, for the judgment of the individual studies it is necessary that the constituents of the individual SJW preparations are clearly described, also with regard to their stability, and that it is demonstrated that adequate clinical trials for the short- and necessary long-term are performed. A cross reference from one SJW extract to another extract is not legitimate, as it is not scientific practice to relate from one SSRI to another without asking for clinical studies and exact pharmacodynamic profile.

References

1 Kessler RC, McGonagle KA, Zhao S, Nelson CB, Hughes M, Eshleman S, Wittchen HU, Kendler KS (1994) Lifetime and 12-month prevalence of DSM-III-R psychiatric disorders in the United States. Results from the National Comorbidity Survey. *Arch Gen Psychiatry* 51: 8–19
2 World Health Organization (1998) Health 21 – the Health for All Policy for the WHO European Region – 21 Targets for the 21st Century (EUR/RC4810). World Health Organization, Geneva
3 Gupta RK, Möller HJ (2003) St. John's Wort. An option for the primary care treatment of depressive patients? *Eur Arch Psychiatry Clin Neurosci* 253: 140–148
4 Kasper S, Trautmann-Sponsel RD (2001) St. John's Wort extract: an effective and well-tolerated antidepressant. *Primary Care Psychiatry* 7: 39–42
5 Laakmann G, Jahn G, Schüle C (2002) *Hypericum perforatum* Extrakt bei der Behandlung leichter bis mittelschwerer Depressionen. Klinische und pharmakologische Aspekte. *Der Nervenarzt* 73: 600–612
6 Linde K, Ramirez G, Mulrow CD, Pauls A, Weidenhammer W, Melchart D (1996) St. John's Wort for depression – an overview and meta-analysis of randomized clinical trials. *BMJ* 313: 253–258
7 Linde K, Mulrow CD (1998) Review: St. John's Wort for depression. *The Cochrane Library* 4: 1–15
8 Linde K, Mulrow CD (2004) *St. John's Wort for depression* (Cochrane Review). The Cochrane Library, Issue 2, 2004. John Wiley and Sons, Ltd, Chichester, UK
9 Maurer A, Johne A, Bauer S et al. (1999) Interaction of St. John's Wort extract with phenprocoumon. *Eur J Clin Pharmacol* 55: A22 (Abstr 79)
10 Schulz V, Hänsel R (1999) *Rationale Phytotherapie. Ratgeber für die ärztliche Praxis*. Springer, Heidelberg
11 Volz HP (1997) Controlled clinical trials of *Hypericum* extracts in depressed patients: An overview. *Pharmacopsychiatry* 30 (Suppl): 72–76
12 Kalb R, Trautmann-Sponsel RD, Kieser M (2001) Efficacy and tolerability of *Hypericum* extract WS® 5572 *versus* placebo in mildly to moderately depressed patients. A randomized double-blind multicenter clinical trial. *Pharmacopsychiatry* 34: 96–103
13 Lecrubier Y, Clerc G, Didi R, Kieser M (2002) Efficacy of St. John's Wort extract WS 5570 in major depression: a double-blind, placebo-controlled trial. *Am J Psychiatry*: 1361–1366
14 Kasper S, Dienel A (2002) Cluster analysis of symptoms during antidepressant treatment with *Hypericum* extract in mildly to moderately depressed out-patients. A meta-analysis of data from three randomized, placebo-controlled trials. *Psychopharmacology* 164: 301–308
15 Shelton RC, Keller MB, Gelenberg A, Dunner DL, Hirschfeld R, Thase ME, Russell J, Lydiard RB, Crits-Cristoph P, Gallop R et al. (2001) Effectiveness of St. John's Wort in major depression. A randomized controlled trial. *JAMA* 285: 1978–1986
16 Davidson JRT on behalf of the Hypericum Depression Trial Study Group (2002) Effect of *Hypericum perforatum* (St. John's Wort) in major depressive disorder. *JAMA* 287: 1807–1814
17 Rychlik R, Siedentop H, von den Driesch V, Kasper S (2001) Johanniskraut WS 5572 bei leichten bis mittelschweren Depressionen. Wirksamkeit und Verträglichkeit von 600 und 1200 mg Wirkstoff täglich. *Fortschritte der Medizin* 119: 119–128
18 Vorbach EU, Arnoldt KH, Mannel M (2000) Johanniskraut als Phyto-Antidepressivum. *Fundamenta Psychiatrica* 14: 172–180

19 Roots I (1999) Arzneimittel-Interaktionen mit Johanniskraut. Gutachten für das BfArM vom 1.12.1999
20 Johne A, Brockmöller J, Bauer S, Maurer A, Langheinrich M, Roots I (1999) Pharmacokinetic interaction of digoxin with an herbal extract from St. John's Wort. *Clin Pharmacol Ther* 66: 338–345
21 Ernst E (1999) Second thoughts about safety of St. John's Wort. *Lancet* 354: 2014–2016
22 Piscitelli SC, Burstein AH, Chaitt D, Alfaro RM, Falloon J (2000) Indinavir concentrations and St. John's Wort. *Lancet* 355: 547–548
23 Laakmann G, Schüle C, Baghai T (1998) St. John's Wort in mild to moderate depression: the relevance of hyperforin for the clinical efficacy. *Pharmacopsychiatry* 31 (Suppl. 1): 54–59
24 Kalb R, Arnoldt KH, Kleemann A (2000) Langzeitbehandlung mit Johanniskraut-Extrakt LI 160. Ergebnisse einer offenen multizentrischen Therapiestudie zur Wirksamkeit und Verträglichkeit über 12 Monate bei Patienten mit depressiven Erkrankungen. *Der Allgemeinarzt* 16: 428–433
25 Kasper S, Dienel A, Kieser M (2004) Continuation and long-term maintenance treatment with *Hypericum* extract WS 5570 after successful acute treatment of mild to moderate depression – Rationale and study design. *Int J Methods Psychiatr Res* 13: 176–183
26 Sommer H, Harrer G (1994) Placebo-controlled double blind study examining the effectiveness of a *Hypericum* preparation in 105 mildly depressed patients. *J Geriatr Psychiatry Neurol* 7 (Suppl 1): S9–S11
27 Hübner WD, Lande S, Podzuweit H (1994) *Hypericum* treatment of mild depressions with somatic symptoms. *J Geriatr Psychiatry Neurol* 7 (Suppl 1): S12–S14
28 Hänsgen KD, Vesper J, Ploch M (1994) Multicentre double-blind study examining the antidepressant effectiveness of the *Hypericum* extract LI 160. *J Geriatr Psychiatry Neurol* 7 (Suppl 1): S15–S18
29 Bjerkenstedt L (2000) Studie mit Johanniskraut-Extrakt LI 160 bei leichten bis mittelschweren Depressionen. Äquipotente Wirkung im Vergleich mit einem synthetischen Antidepressivum gezeigt. *Ärzte Zeitung* 18 (März), Sonderdruck Nr. 226
30 Montgomery SA, Hübner WD, Grigoleit HG (2000) Efficacy and tolerability of St. John's Wort extract compared with placebo in patients with a mild to moderate depressive disorder. *Phytomedicine* 7 (Suppl II): 107
31 Schrader E, Meier B, Brattström A (1998) *Hypericum* treatment of mild-moderate depression in a placebo-controlled study. A prospective, double-blind, randomized, placebo-controlled, multicentre study. *Hum Psychopharmacol* 13: 163–169
32 Philipp M, Kohnen R, Hiller KO (1999) *Hypericum* extract *versus* Imipramine or placebo in patients with moderate depression. Randomised multicenter study of treatment for eight weeks. *Brit Med J* 319: 1534–1538
33 Harrer G, Schmidt U, Kuhn U, Biller A (1999) Comparison of equivalence between St. John's Wort extract LoHyp-57 and Fluoxetine. *Arzneimittel-Forschung/Drug Research* 49: 289–296
34 Schrader E (2000) Equivalence of St. John's Wort extract (ZE 117) and fluoxetine: a randomized, controlled study in mild-moderate depression. *Int Clin Psychopharmacol* 15: 61–68
35 Woelk H (2000) Comparison of St. John's Wort and imipramine for treating depression: randomised controlled trial: *Brit Med J* 321: 536–539
36 Volz HP, Laux P (2000) Potential treatment for subthreshold and mild depression: a comparison of St. John's Wort extracts and fluoxetine. *Compr Psychiatry* 41 (Suppl 1): 133–137

Clinical efficacy of St. John's Wort in psychiatric disorders other than depression

Hans-Peter Volz

Krankenhaus für Psychiatrie und Psychotherapie Schloss Werneck, Balthasar-Neumann-Platz 1, D-97440 Werneck, Germany

St. John's Wort (*Hypericum*) extracts are among the leading antidepressants in Europe [1]. Although some of the clinical trials have been challenged because of questionable methodology [2], it is clear that these extracts are more efficacious than placebo and as efficacious as synthetic antidepressants in treating mild-to-moderate depressive episodes (e.g., [3–7]). However, data concerning the efficacy of St. John's Wort in psychiatric disorders other than depression are sparse. That is astonishing in so far as there is good evidence for efficacy in animal models of anxiety (e.g., [8–11]). It has been speculated that hyperforin might be the compound of *Hypericum* extracts possessing the decisive pharmacological anxiolytic property (e.g., [12]).

Apart from the preclinical data indicating an efficacy of *Hypericum* extracts in areas other than depression, some clinical trials show a broad efficacy profile, too.

An interesting approach was chosen by Kasper and Dienel [2]. They reanalysed data of three randomised, placebo-controlled trials with the *Hypericum* extracts WS 5570 and WS 5572 comprising a total of 544 mildly-to-moderately depressed patients. The primary outcome criterion was the score on the Hamilton-Depression-Scale, 17 items (HAMD). In all three investigations, treatment with the *Hypericum* extract demonstrated a statistically significant superiority over placebo. The results of the individual HAMD-items were used to perform a hierarchical cluster analysis. This analysis revealed two clusters: Cluster I comprised what the authors' named "core-symptoms" of depression (item 1: depressed mood, 2: guilt, 3: suicide, 7: work and interest, 8: retardation, 12: gastrointestinal symptoms, 13: general somatic symptoms, 14: genital symptoms, 16: loss of weight); cluster II comprised the remaining depression-related symptoms, namely sleep disturbances and anxiety (item 4: insomnia, initial, 5: insomnia, middle, 6: insomnia, delayed, 9: agitation, 10: anxiety, psychic, 11: anxiety, somatic, 15: hypochondriasis, 17: loss of insight).

Separate analysis of the score-reduction in Clusters I and II (Fig. 1) showed that the sequence of change for Clusters I and II was very similar, with only

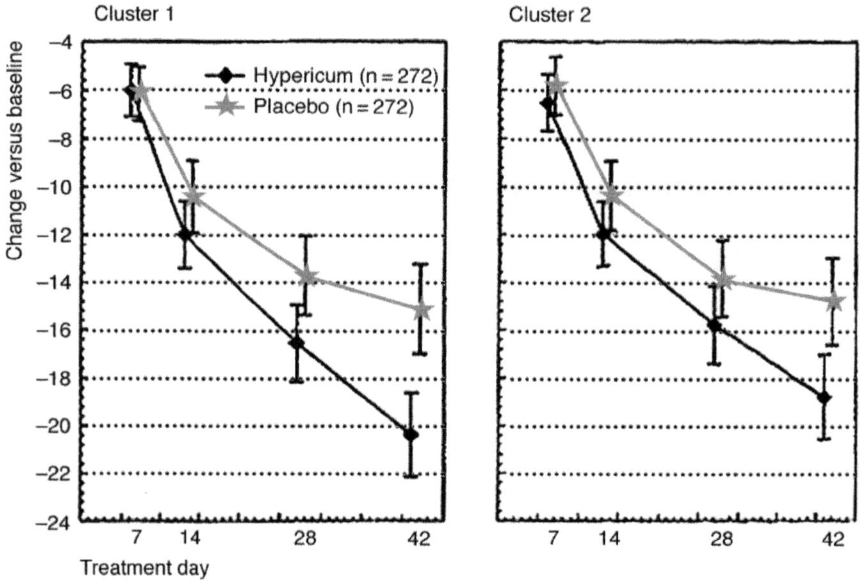

Figure 1. Rescaled cluster scores. The change from baseline is given (Hamilton Depression Scale, 17-items, rescaled cluster, see text) between baseline and subsequent visits (mean and 95% confidence interval, negative values indicate symptom improvement) (according to [2]).

slightly more pronounced score decreases for Cluster I than Cluster II. Both clusters demonstrated notably more pronounced average score decreases in patients treated with the *Hypericum* extract as compared to placebo, especially during the second half of the treatment period, when the curves of the placebo group flattened out whereas those in the *Hypericum* extract group still showed a stable score decrease. For Cluster I the 50% responder-rates (percentage of patients showing at least a 50% decrease in the sum score) were 56.6% for the *Hypericum* extract patients (41.9% for the placebo patients), for Cluster II 47.8% (34.2%). The most pronounced single-item differences (Fig. 2) between *Hypericum* extract and placebo were found for item 1 ("depressed mood"), 8 ("retardation"), 13 ("general somatic symptoms", 9 and 14 ("genital symptoms") (Cluster I) and for item 10 ("psychic anxiety") (Cluster II). The authors conclude that "as regards the therapeutic profile of *Hypericum* extract, the drug appears to act primarily on the core symptoms of depression, with additional beneficial effects on accompanying, depession-related anxiety".

The analysis shows that depression-related anxiety, which also includes somatic anxiety, responds very well to *Hypericum* extract. However, with other single items related to somatic complaints like "general somatic symptoms", "genital symptoms" and "gastrointestinal symptoms", *Hypericum* treatment also took a very pronounced effect. Besides the depressive symptomatology

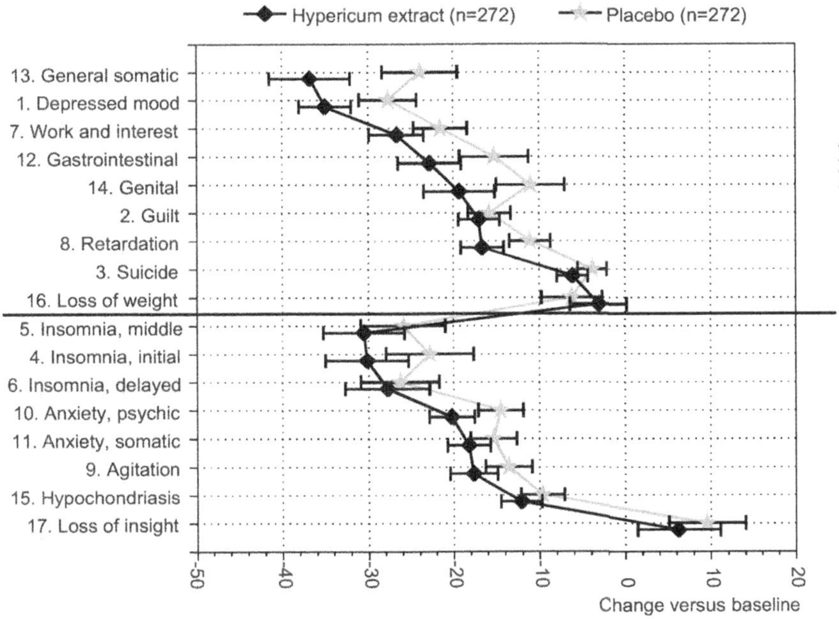

Figure 2. Rescaled item score change *versus* baseline (mean and 95% confidence interval, negative values indicate symptom improvement, items ordered within clusters by decreasing magnitude of change in the *Hypericum* group) (according to [2]).

the somatic complains responded best. These findings, based on a cluster analysis, were corroborated by results from a double-blind study reporting efficacy of the *Hypericum* extract LI 160 on headache and on cardiac symptoms in patients with neurotic depression or brief depressive reactions [13].

As will be shown below, this is also true for somatic complaints in patients not primarily depressed.

In an early, small-scale, placebo-controlled, randomised study (n = 39) [14] including depressive patients with somatic symptoms, a similar trend was found. Not only was the depressive symptomatology reduced, but the accompanying somatic symptoms also showed a good treatment effect: At baseline 16 patients in the *Hypericum* group complained of palpitations as compared to 17 in the placebo group. The respective numbers after a 4-week treatment period were 8 and 12. For headache, the baseline frequency distribution was 9/11, at week 4 it was 2/7; for muscle pain the difference was even more pronounced, 8/12 at baseline and 2/10 after 4 weeks.

These early findings of the efficacy on somatic symptoms led to the decision to perform a placebo-controlled trial in patients suffering from somatoform disorders [15]. These disorders represent a group of diseases that is characterised by physical complaints that lead the patient to believe that he or she suffers from a physical disease; however, no sufficient diagnosis of a known medical condition can be found despite thorough investigation. The patients seek medical help, and, typically, they fail to believe that their complaints possess a psychic background. They often see several physicians, and thus cause a high economic burden. Recently, it has been stressed that somatoform disorders represent severe psychiatric illnesses which are widely neglected by psychiatrists [16]. One problem of the disease concept consists in a potential diagnostic overlap with depression and anxiety disorder. In a 6-week trial the efficacy and tolerability of the St. John's Wort extract LI 160 (600 mg/die) was compared to placebo in 151 outpatients suffering from somatisation disorder (ICD-10: F45.0), undifferentiated somatoform disorder (F45.1), or somatoform autonomic dysfunction (F45.3) (for the demographic data, see Tab. 1). As the main outcome criterion the subscale "somatic anxiety" from the Hamilton-Anxiety-Rating-Scale (HAMA) was chosen. The HAMA-SOM consists of the following items: 7 (somatic [muscular] symptoms), 8 (somatic [sensory] symptoms), 9 (cardiovascular symptoms), 10 (respiratory symptoms), 11 (gastrointestinal symptoms), 12 (genitourinary symptoms), and 13 (autonomic symptoms). By using these criteria, HAMA-SOM reflects most of the complaints of patients with somatoform disorder. As regards the results, LI 160 turned out to be superior concerning the primary outcome criterion HAMA-SOM (decrease from 15.39 [SD: 2.68] to 6.64 [4.32] in the *Hypericum* group and from 15.55 [2.94] to 11.97 [5.58] in the placebo group [statistically significant difference, $p = 0.001$]) (Fig. 3, Tab. 2). This was corroborated by the result of a statistically significant superior efficacy in the outcome criteria additionally used such as Clinical Global Impression (CGI), HAMA-total score, HAMA, subscore psychic anxiety, Hamilton Depression Scale, Self-Report Symptom Inventory 90 Items-revised (SCL-90-R), and

Table 1. Demographic data of the included patients (intent-to-treat population). The diagnoses were fixed at baseline. The total amount of diagnoses is larger than the number of patients due to double diagnoses in some patients (according to [15]).

	Hypericum	Placebo
Total number	75	74
Female, number (%)	54 (72)	46 (62)
Male, number (%)	21 (28)	28 (38)
Mean age, years (SD)	46.9 (12.2)	48.6 (11.8)
Somatisation disorder (F45.0), number (%)	55	53
Undifferentiated somatisation disorder (F45.1), number (%)	2	5
Somatoform autonomic disorder (F45.3), number (%)	19	19

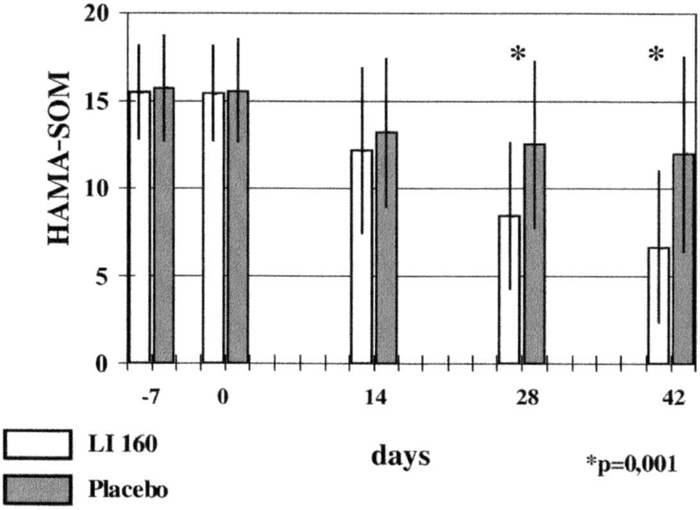

Figure 3. Subscore "somatic anxiety" of the Hamiliton Anxiety Scale (HAMA-SOM) during the 6 week double-blind treatment period. Mean and standard deviation are given for the intent to treat (ITT) population (*Hypericum* n = 75, placebo n = 74).

Table 2. Summary of results of the ITT-efficacy analyses: mean scores of scales (standard deviations in brackets), p-values for univariate analysis of covariance (according to [15])

	day 0		day 42		p
	Hypericum	Placebo	*Hypericum*	Placebo	
HAMA-SOM	15.39 (2.68)	15.55 (2.94)	6.64 (4.32)	11.97 (5.58)	.001*
HAMA	22.09 (4.65)	22.47 (5.01)	10.00 (6.39)	17.00 (7.71)	.0001**
HAMA-PSY	6.71 (2.82)	6.92 (2.79)	3.36 (2.86)	5.03 (3.16)	.0001
HAMD	10.59 (2.40)	10.80 (2.74)	5.43 (3.85)	8.08 (4.10)	.0001#
SCL-90-R	61.65 (43.56)	66.37 (45.61)	29.38 (33.70)	50.50 (38.85)	.0001##
SCL-90-R-SOM	15.57 (8.48)	15.95 (7.32)	6.84 (5.53)	12.50 (7.63)	.0001+

(Abbreviations: HAMA = Hamilton Anxiety Scale, HAMA-SOM = subfactor somatic anxiety of the HAMA, HAMA-PSY = subfactor psychic anxiety of the HAMA, HAMD = Hamilton Depression Scale, SCL-90-R = Self-Report Symptom Inventory 90 Items-Revised, SCL-90-R-SOM = somatic subfactor of the SCL-90-R).

* Difference reached statistical significance (p = 0.001) on day 28.

** Difference reached statistical significance (p = 0.0001) on day 28.

\# Difference reached statistical significance (p = 0.0012) on day 28.

\## Difference reached statistical significance (p = 0.034) on day 14, on day 28 p-value was p = 0.0001.

\+ Difference reached statistical significance (p = 0.0001) on day 28.

SCL-90-R, subscore somatic anxiety. In order to investigate the potential influence of the reduction of depressive symptomatology on the outcome of the main parameter, the included population was split into those patients with no and those with mild depressive symptoms. In both subpopulations, LI 160 took effect on the somatoform symptoms, showing that its efficacy is independent of an existing depressive mood. Tolerability of LI 160 was excellent, which was shown both by the number of adverse events being identical in both groups and by the overall tolerability statement of the investigators and the patients.

A similar study was performed by Müller et al. [17]. 175 patients suffering from somatoform disorder were included for a period of 6 weeks. They received in a randomised and double-blind fashion either 300 mg LI 160 b.i.d. or matching placebo. Six outcome criteria (Somatoform Disorders Screening Instrument – 7 days [SOMS-7], somatic subscore of the HAMA, somatic subscore of the SCL-90-R, subscores "improvement" and "efficacy" of the CGI, global judgement of efficacy by the patient) were evaluated as a combined measure by means of the Wei Lachin test. LI 160 was statistically highly significantly superior effective than placebo in the combined efficacy parameter ($p < .0001$) and in the *post hoc* analysed single efficacy parameters (Tab. 3). Of the LI 160 patients, 45.4% were rated as responders (defined as a decrease ≥ 50% of the SOMS-7-score and a "very much better" or "much better" rating on the CGI "improvement" scale at week 6), compared to 20.9% of the placebo-patients ($p = .0006$). Tolerability of LI 160 was equivalent to placebo.

Thus, two positive placebo controlled trials showing superior efficacy and excellent tolerability of LI 160 in somatoform disorders are available.

In two studies the effect of St. John's Wort extracts on premenstrual respectively menopausal symptoms is reported [18, 19].

The premenstrual syndrome is characterised by physical, behavioural and psychological symptoms appearing regularly during the week prior to menstruation and disappearing within a few days of onset, causing disruption of occupational, family and personal functioning (cited according to [18]). The

Table 3. Summary of results of the ITT-efficacy (last value carried forward) analyses of the trial of Müller et al. [17]. Shown are mean scores of scales (standard deviations in brackets) at baseline and after 6 weeks of treatment, p-values for the one-sides Wei-Lachin test ($\alpha = 2.5\%$).

	day 0		day 42		p
	Hypericum	Placebo	*Hypericum*	Placebo	
HAMA-SOM	12.8 (2.3)	12.7 (2.5)	5.8 (4.2)	8.4 (4.6)	<.0001
SCL-90-R-SOM	15.8 (6.2)	15.4 (5.9)	8.5 (6.3)	12.5 (6.7)	<.0001
SOMS-7	23.2 (4.2)	23.5 (5.5)	12.4 (8.6)	18.3 (9.4)	<.0001

Abbreviations: HAMA-SOM = subfactor somatic anxiety of the Hamilton Anxiety Scale, SCL-90-R-SOM = somatic subfactor of the Self-Report Symptom Inventory 90 Items-Revised, SOMS-7 = Somatoform Disorders Screening Instrument – 7 days.

prevalence is estimated to be 2–5% among women of childbearing age. The authors of this open trial investigated four cycles in a total of 19 women. Within the treatment period of two cycles the patients received one 300 mg tablet/day of a *Hypericum* extract standardised to 900 µg hypericin (Kira® by Lichtwer Pharma). The symptoms decreased by approximately 50%. However, since the placebo response rate in other trials in premenstrual syndrome were even higher than 50%, it is impossible to decide how much of the effect was attributable to the phytopharmacon. Nevertheless, for the authors these results seem promising enough to ask for the performance of controlled clinical trials.

Grube et al. investigated the effect of St. John's Wort extract (3 × 135–225 mg standardised to a total hypericin content of 900 µg) on menopausal symptoms [19]. Such menopausal or climacteric symptoms can, according to the authors, be characterised as psychological (irritability, disturbed sleep, inner tension, anxiety, depressive mood, subjectively perceived decrease in physical attractiveness, decrease in libido), psychosomatic or vegetative (hot flushes, outbreaks of sweating, palpitations, dizziness, headaches) and organic (skin and mucosal atrophy, osteoporosis, and arteriosclerosis) symptoms. Treatment with estrogens is the standard therapy. However, considering the possible long-term side effects of this treatment (e.g., hypertension, body weight increase, gallstone formation), this replacement therapy is not unreservedly recommendable. As an alternative treatment, St. John's Wort extract was investigated in a total of 111 women within a 12-week treatment period. In all investigated syndromes symptomatology was clearly reduced (Fig. 4). Regarding sexual complaints, the symptom reduction was most

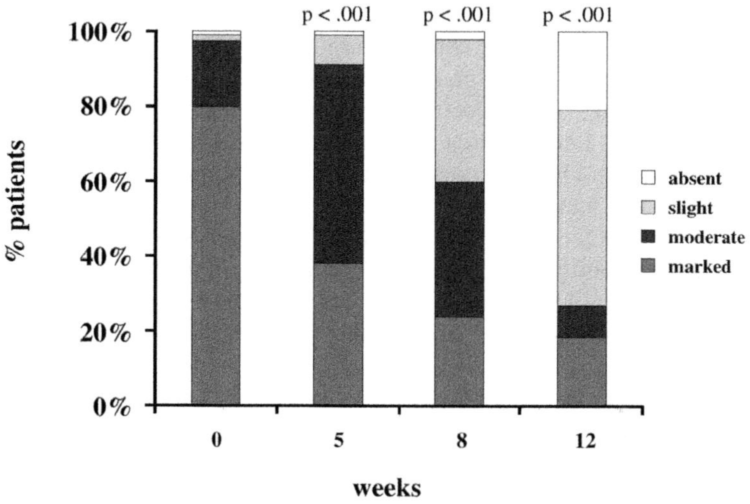

Figure 4. Incidence and severity of outbreaks of sweating during treatment (modified according to [19]).

impressive. However, since no placebo control was included in this trial, positive results remain preliminary, and placebo controlled trials to prove efficacy are needed.

A small open trial studied the effect of St. John's Wort extract (3 × 135–225 mg standardised to a total of 900 μg hypericin) on 20 patients with a fatigue syndrome [20]. However, inclusion of patients did not follow strict diagnostic guidelines. Inclusion criteria were complaints of fatigue, tiredness, and exhaustion without an overt medical reason other than depression. Only those patients who replied in the negative when asked "do you believe you are depressed?" were included. Although patients had thus denied being depressed, half of the sample was in fact depressed according to the scores of the Hospital Anxiety and Depression (HAD) scale. In the course of the trial the patients exhibited a decrease in their fatigue. Depression and anxiety scores also decreased significantly from baseline to the end of the study at week 6. Patients suffering from depressive symptoms at the start of the trial were significantly more likely to have an improvement in fatigue than those who were not depressed. This improvement in fatigue might be an indirect result of the (specific?) effects of the St. John's Wort extract on depression, with the perceived reduction in fatigue reflecting only a single symptom of the whole spectrum of depressive symptomatology. Due to the open design of the study, but also due to the problem just mentioned, controlled data are required on depressed patients with a fatigue syndrome and on non-depressed patients with a fatigue syndrome in order to be able to isolate the potential effect of St. John's Wort on fatigue *per se*.

Regarding the syndrome fatigue/somatisation, Murck discusses these disorders in the context of the atypical depression (AD) spectrum [21]. He claims that major depression is a heterogeneous group of disorders, with typical depression on the one side, and AD on the other. According to DSM-IV AD is characterised by mood reactivity (in typical depression, this reactivity is decreased) and so-called "reversed" vegetative symptoms such as significant weight gain or increase in appetite, hypersomnia, leaden paralysis and long-standing pattern of interpersonal rejection sensitivity. Based on similarities in symptomatology and high comorbidity rates Murck points to the AD spectrum consisting of AD and chronic fatigue syndrome, somatoform disorder and fibromyalgia. A potential common physiological basis of the AD spectrum might be, as in Cushing's disease, a reduced activity of the hypothalamus–pituitary–adrenocortical (HPA)-system (in typical depression the activity of the HPA-system is increased), an activity that can be determined by measuring plasma and urine cortisol levels.

As regards the treatment of the AD spectrum, Murck argues that *Hypericum* extracts might be of special value: A study in patients with mild-to-moderate depression compared the effect of *Hypericum* extract LI 160 with fluoxetine and placebo [22] and found a significant superiority of *Hypericum* over fluoxetine and a trend to superiority over placebo. As Murck mentioned [21], a subgroup analysis including patients with reversed vegetative signs was performed, showing, interestingly, the same pattern in this subgroup with a much more pronounced effect size, but no efficacy of any drug compared with placebo in the non-atypical patients [22]. In another subgroup of depression with fatigue as one main symptom and reversed vegetative symptom as the other main characteristic, seasonal affective disorder (SAD), *Hypericum* extracts showed efficacy in open studies [23–25].

As regards anxiety disorders, only one open study and several case reports have been published so far.

One open trial [26] described the treatment effects of St. John's Wort on 12 subjects suffering from obsessive–compulsive disorder (OCD) according to DSM-IV-criteria. The authors give no clear-cut information on whether only hypericin or an *Hypericum* extract containing hypericin was used, stating that the treatment was performed "with a fixed dose of 450 mg of 0.3% hypericin (a psychoactive compound in *Hypericum*) twice daily (extended-release formulation)". The treatment lasted for 12 weeks. The score of the Yale-Brown Obsessive Compulsive Scale (Y-BOCS) dropped by 7.42 points (statistically significant decrease from baseline). The decrease was significant in each individual patient. At endpoint, 42% of the patients rated themselves as "much" or "very much improved" on the PGI (Patients Global Impressions of Improvement). Given that six patients pretreated with selective serotonin reuptake inhibitor (SSRIs) reported a failure to respond, this group was investigated separately. Those patients had a mean Y-BOCS change of 4.33 points, whereas those who did not report previous treatment failure had a Y-BOCS score change of 10.50 points. The authors' emphasise the fact that the treatment-effect observed in this open trial was similar to an SSRI-effect. In the light of these promising results, the authors propose to perform a randomised, placebo controlled trial in order to further investigate the potential of St. John's Wort in reducing obsessive–compulsive symptoms.

Davidson and Connor [27] report on three at least partially treatment-resistant patients suffering from generalised anxiety disorder (GAD). The patients were treated with 1,500 mg/day (one case) or 1,800 mg/day (two cases) of St. John's Wort extract. The treatment results were very good, and the authors propose to further assess the efficacy of St. John's Wort in GAD with special regard to the value of high daily doses.

Conclusion

The best results for St. John's Wort extracts in psychiatric indications other than depression have been measured for somatoform disorders, which might be linked to the spectrum atypical depression/fatigue as proposed by Murck [21]. Only very preliminary data are available for anxiety disorders, although St. John's Wort extracts seem to be nearly as efficacious in anxiety symptoms (in depressed patients) as in core-symptoms of depression.

References

1 Harrison P (1998) Herbal medicine takes root in Germany. *CMAJ* 158: 637–639
2 Kasper S, Dienel A (2002) Cluster analysis of symptoms during antidepressant treatment with *Hypericum* extract in mildly to moderately depressed out-patients. A meta-analysis of data from three randomised, placebo-controlled trials. *Psychopharmacology* 164: 301–308
3 Linde K, Ramirez G, Mulrow CD, Pauls A, Weidenhammer W, Mechart D (1996) St. John's Wort for depression – an overview and meta-analysis of randomised controlled clinical trials. *BMJ* 313: 253–258
4 Volz HP (1997) Controlled clinical trials of *Hypericum* extracts in depressed patients – an overview. *Pharmacopsychiatry* 30: 102–107
5 Kim HL, Streltzer J, Goebert D (1999) St. John's Wort for depression – a meta-analysis of well-defined clinical trials. *J Nerv Ment Dis* 187: 532–539
6 Kasper S, Trautmann-Sponsel RD (2001) St. John's Wort extract: an effective and well-tolerated antidepressant. *Primary Care Psychiatry* 7: 39–42
7 Kasper S (2004) Clinical efficacy in depression. In: WE Müller (ed.): *St. John's Wort and its active principles in depression and anxiety*. Birkhäuser, Basel; 123–131
8 Beijamini V, Andreatini R (2003a) Effects of *Hypericum perforatum* and paroxetine on rat performance in the elevated T-maze. *Pharmacol Res* 48: 199–207
9 Beijamini V, Andreatini R (2003b) Effects of *Hypericum perforatum* and paroxetine in the mouse defense test battery. *Pharmacol Biochem Behav* 74: 1015–1024
10 Flausino OA Jr, Zangrossi H Jr, Salgado JV, Viana MV (2002) Effects of acute and chronic treatment with *Hypericum perforatum* L. (LI 160) on different anxiety-related responses in rats. *Pharmacol Biochem Behav* 71: 251–257
11 Chatterjee SS (2004) Efficacy in behavioural models of anxiety and cognition. In: WE Müller (ed.): *St. John's Wort and its active principles in depression and anxiety*. Birkhäuser, Basel; 85–97
12 Zanoli P, Rivasi M, Baraldi C, Baraldi M (2002) Pharmacological activity of hyperforin acetate in rats. *Behav Pharmacol* 13: 645–651
13 Sommer H, Harrer G (1994) Placebo-controlled double-blind study examining the effectiveness of a *Hypericum* preparation in 105 mildly depressed patients. *J Geriatr Psychiatry Neurol* 7: S9–S11
14 Hübner WD, Lande S, Podzuweit H (1994) *Hypericum* treatment of mild depressions with somatic symptoms. *J Geriatr Psychiatr Neurol* 7 (Suppl 1): S12–S14
15 Volz HP, Murck H, Kasper S, Möller HJ (2002) St. John's Wort extract (LI 160) in somatoform disorders: results of a placebo-controlled trial. *Psychopharmacology* 164: 294–300
16 Bass C, Peveler R, House A (2001) Somatoform disorders: severe psychiatric illnesses neglected by psychiatrists. *Br J Psychiatry* 179: 11–14
17 Müller T, Mannel M, Murck H, Rahlfs VW (2004) Treatment of somatoform disorders with St. John's Wort – a randomised, double-blind and placebo-controlled trial. *Psychosom Med* 66: 538–547
18 Stevinson C, Ernst E (2000) A pilot study of *Hypericum perforatum* for the treatment of premenstrual syndrome. *Br J Obstet Gynaecol* 107: 870–876
19 Grube B, Walper A, Wheatley D (1999) St. John's Wort extract: Efficacy for menopausal symptoms of psychological origin. *Advances in Therapy* 16: 177–1186
20 Stevinson C, Dixon M, Ernst E (1998) *Hypericum* for fatigue – a pilot study. *Phytomedicine* 5: 443–447
21 Murck H (2003) Atypical depression and spectrum disorder – neurobiology and treatment. *Acta Neuropsychiatrica* 15: 227–241
22 Murck H, Fava M, Alpert J, Nierenberg AA, Mischoulon D, Otto MW, Zajecka J, Mannel M, Rosenbaum JF (2004) *Hypericum* extract in patients with MDD and reversed vegetative signs: re-analysis from data of a double-blind, randomized trial of *Hypericum* extract, fluoxetine and placebo. *Int J Neuropsychopharmacol* sept. 30, 1–7 (Epub ahead of print)
23 Wheatley D (1999) *Hypericum* in seasonal affective disorder (SAD). *Curr Med Res Opinion* 15: 33–37
24 Kasper S (1997) Treatment of seasonal affective disorder (SAD) with *Hypericum* extract. *Pharmacopsychiatry* 30 (Suppl 2): 89–93
25 Martinez B, Kasper S, Ruhrmann S, Möller HJ (1994) *Hypericum* in the treatment of seasonal

affective disorders. *J Geriatr Psychiatry Neurol* 7, Suppl. 1: S29–S33
26 Taylor LH, Kobak KA (2000) An open-label trial of St. John's Wort (*Hypericum perforatum*) in obsessive-compulsive disorder. *J Clin Psychiatry* 61: 575–578
27 Davidson JRT, Connor KM (2001) St. John's Wort in generalized anxiety disorder: Three case reports. *J Clin Psychopharmacol* 21: 635–636

St. John's Wort and its Active Principles in Depression and Anxiety
Edited by W. E. Müller

Side effects and drug interactions

Volker Schulz[1] and Andreas Johne[2]

[1] *Oranienburger Chaussee 25, 13465 Berlin, Germany,* [2] *Institut für Klinische Pharmakologie, Universitätsklinikum Charite, Schumannstraße 20, 10098 Berlin, Germany*

Incidence of adverse events during clinical trials

With the exception of objectively measurable side effects and interactions, adverse events (AE) observed during clinical trials were predominantly associated with the setting of the study and, where applicable, the comparator medication. Accordingly, the incidence of subjective complaints varied from study to study between 0% and almost 100%. As a rule, clinical trials carried out with healthy subjects had the highest and placebo controlled studies with ambulatory patients of family physicians the lowest rate of adverse events [1–5]. Subjective events such as gastric complaints, itching, tiredness, sleep disorders and headaches that were recorded in the active treatment group of placebo controlled observational studies can most likely be causally attributed to the *Hypericum* extracts. Objectively measurable AE during clinical trials, as far as photosensitisation and drug interactions are concerned, will be discussed below. An actual overview on side effects in all clinical studies with St John's Wort is given by Knüppel and Linde [5].

Incidence of adverse events in observational studies

For reasons of numbers alone, observational studies are more suitable than clinical trials for the qualitative and quantitative assessment of subjective adverse events. Observational studies are now available for drug products containing St John's Wort dry extracts prepared with alcohol (80% methanol or 60% ethanol in water as the extraction agent) in more than 10,000 patients [5, 6]. The incidence of adverse events among patients was 1–3%. Similar results were seen in three observational studies undertaken with Jarsin® in a total of 5,700 patients. The first study [7] involved 3,250 patients of whom 76% were women and 24% men. The age of patients ranged from 20–90 years with an average of 51 years. 49% of the patients had mild, 46% moderately severe and 3% severe depression. Adverse events were reported by 79 (2.43%) and treatment withdrawals by 48 (1.45%) of patients. The most frequent side effects were gastrointestinal symptoms, allergic reactions, tiredness and restlessness (Tab. 1).

Table 1. Adverse events (AE) among 3,250 patients treated with the commercial product Jarsin® [7]

Spontaneously reported AE	
Gastrointestinal symptoms: Nausea 6, hypogastric pain 5, loss of appetite 3, diarrhoea 2, gastric complaints 2	18 (0.55%)
Allergic reactions: allergy 6, skin rash 6, pruritus 5	17 (0.52%)
Tiredness	13 (0.40%)
Anxiety	8 (0.26%)
Confusion	5 (0.15%)
Other ADR*	18 (0.55%)
Total number of ADR	79 (2.43%)

* Under "other side effects", two patients each reported dry mouth, sleep disorders, palpitations, weakness and worsening of concurrent diseases. Heart flutter, circulatory complaints, irritability, visual disorders, disorders of micturition, burning eyes, euphoria and nervous tension were each reported once.

Case reports from pharmacovigilance

Between October 1991 and December 1999, some 8.6 million patients are estimated to have been treated with Jarsin®, the leading *Hypericum* preparation at that time in Germany. Over this period, the German adverse drug reaction (ADR) recording system received 95 spontaneous reports of side effects for this product (Tab. 2). Top of the list were 27 reports of "allergic" skin reactions, 16 reports of increases in Quick values (prothrombin time) under concomitant treatment with coumarin-type anticoagulants, nine of gastrointestinal disorders, eight of breakthrough bleeding during concomitant ingestion of

Table 2. Treated cases and spontaneously reported AE under treatment with Jarsin® [8]

Year	Number of tablets sold	Number of patients estimated*	ADR reports
1991	2,455,000	19,000	-
1992	32,535,000	250,000	1
1993	45,061,500	350,000	3
1994	77,332,350	600,000	6
1995	140,414,480	1,100,000	11
1996	200,009,760	1,570,000	11
1997	192,744,220	1,530,000	8
1998	225,283,500	1,787,000	23
1999	174,145,350	1,382,000	32
1991–1999		8,588,000	95

* Basis of calculation: 6 weeks' treatment in each case, daily dose 3×1 tablet; Number of patients = number of tablets sold divided by 126.

contraceptives, seven of a reduction in cyclosporin levels in organ transplant recipients, four of tingling paraesthesias and three reports of cardiovascular complaints. Other ADR that were reported less than three times, are shown in Table 3. Severity was rated as mild to moderate in all cases. The side effects were spontaneously reversible without additional therapy when treatment was discontinued and were never dangerous or life-threatening for the patient [8]. The relative frequency of the ADR is shown in Table 3. There were a total of about 1.1 reports per 100,000 patients treated. With roughly one report per 300,000 cases treated, reversible skin reactions (meaning photosensitisation) are the most common. However, severe phototoxic reactions such as the hypericism seen in grazing animals [9, 10] have never been reported.

Of greater relevance for the therapeutic safety of specific groups of patients are the published interactions with phenprocoumon and cyclosporin. These carry the risk of severe complications. However, co-medication with these drugs is only likely with a small proportion of patients, which is why the calculated relative frequency here assumes a far higher importance. The ADR database of the World Health Organization (WHO) Collaborating Centre for International Drug Monitoring comprised a total of 70 case reports from eight countries classified as drug interactions with SJW containing preparations as of June 2003. Records from Austria, Belgium, Spain and the UK were not part of the WHO listing, since the drug safety bodies of these countries do not allow WHO to release case reports. Therefore, reports from the UK (n = 41) were directly received from the Medicines and Health Products Regulatory

Table 3. Spontaneously reported AE under treatment with Jarsin® [8]

Adverse events	Reports	Reports per 100,000 patients
All events	95	1.10
Skin reactions	27	0.31
Quick (prothrombin time) ↑	16	0.19
Gastrointestinal complaints	9	0.10
Breakthrough bleeding (pill)	8	0.09
Plasma cyclosporin ↓	7	0.08
Tingling paraesthesias	4	0.05
Cardiovascular symptoms	3	0.04
Visual disturbances/pain in the eyes, increased skin sensitivity to pain, sleep disorders	each 2	each 0.02
Dizziness, hair loss, tiredness, restlessness/euphoria, fever, dyspnoea, increase in tinnitus, headaches, elevated thyroid hormones, acute conjunctivitis, increased melanin spots, cutaneous Lupus erythematosus, pain under the scalp, metallic taste, galactorrhoea.	each 1	each 0.01

Agency (MHRA) Yellow Card Scheme as of June 2003. Sources as well as proportions of reports largely vary over time and from country to country, and may be influenced by the extent of use of SJW products, publicity, nature of reactions and other factors. Moreover, no information is provided on the number of patients exposed to SJW products. In the vast majority of cases, sufficient information to reliably estimate interaction likelihood is lacking (e.g., in only three cases outcome of dechallenge and rechallenge are reported), and most cases would have to be classified as "non appraisable" according to a 10-point scoring system recently suggested by Fugh-Berman and Ernst [11]. Thus, information from these reports must be assessed cautiously [12]. Pharmacodynamic interactions with antidepressants (mainly serotonin syndrome, hypertension), and pharmacokinetic interactions with hormonal contraceptives, anticoagulants and immunosuppressants account for 73% of reports [12].

Except for one record of phototoxicity after co-administration of SJW and delta-amino-laevulinic acid for photodiagnosis [13], reports of pharmacodynamic interactions all pertain to centrally acting agents, with serotonin syndrome-like symptoms being the most prominent adverse reactions, when SJW and other serotonergic agents like selective serotonin reuptake inhibitors (SSRIs) were combined. Nearly all pharmacokinetic interactions reported resulted in reduced blood levels or reduced therapeutic response of co-medications. Solely for anticoagulants, in seven of 25 reports an increase in drug effect was observed.

Case reports and clinical trials concerning the phototoxicity

With reference to the disease pattern of hypericism in grazing animals [9, 10] the German monograph "*Hyperici herba*" [14] specifically warns against photosensitising skin reactions. Although comparably serious clinical symptoms had never been seen in humans during treatment with the *Hypericum* extract preparation LI 160, specific investigations were undertaken in healthy subjects to determine the threshold dose of a clinically tolerable photosensitisation in humans [15, 16]. These studies showed that on single use of 1,800 mg or 3,600 mg (containing about 5.5 mg or 11 mg of total hypericin, respectively), only marginal effects on UV-A and UV-B light induced pigmentation occurred. After a daily dose of 1,800 mg for 14 days in one study, the so-called minimal erythema dose (MED) and minimal tanning doses (MTD) were reduced to an extent that would be compensated by a 21% reduction in irradiation of UV-A and solar simulated light [15]. In the second study, no significant effect of visible light and solar simulated light on erythema dose or pigmentation could be detected after a daily dose of 900 mg for 7 days [16]. Therefore, the results of both studies, summarised in Table 4, do not provide evidence for a phototoxic potential of *Hypericum* extract in humans when administered orally in typical clinical doses up to 1,800 mg daily.

Table 4. Effect of *Hypericum* extract on skin sensitivity to ultraviolet A (UVA), ultraviolet B (UVB), visible light (VIS) and solar simulated radiation (SIM) after oral ingestion in healthy subjects

First author, year	N	Dose/day	Duration	Result
Brockmöller, SIM; 1997 [15]	13	Placebo 900 mg 1,800 mg 3,600 mg	1 day	MED unchanged with UVA, UVB and MTD slightly decreased with UVA under 3,600 mg.
Brockmöller, 1997 [15]	50	1,800 mg	14 days	Day 15: MED and MTD slightly decreased with UVA and SIM (could be compensated by 21% reduction in irradiation).
Schempp, 2003 [16]	48	1,800 mg 3,600 mg	1 day	No significant effect of UVA, UVB, VIS and SIM; only marginal effect ($p = 0.05$) on UVB induced pigmentation.
Schempp, 2003 [16]	24	900 mg	7 days	No significant effect of VIS and SIM on MED and MTD.

MED = minimal erythema dose; MTD = minimal tanning dose.

Two cases of severe phototoxic or photoallergic reactions during ingestion of *Hypericum* preparations have been reported in the literature. Acute neuropathy, reversible after discontinuation of the product, occurred in a 35 year-old woman after she had taken 500 mg/day of powdered St. John's Wort (whole herbal drug) for 4 weeks, followed by exposure to sunlight [17]. An itchy rash developed on areas of the skin exposed to light in a 61 year-old woman after ingestion of a St. John's Wort preparation (Hyperforat tablets 3×2 daily) for 3 years. On discontinuation of the preparation, the rash disappeared completely within 2 weeks [18]. However, the doses used – in one case 240 mg and in the other the equivalent dose of 100 mg of extract per day (Tab. 5) – raise doubts concerning the causality of the symptoms in relation to the amount of hypericins ingested with these doses [15].

Table 5. Phototoxicity of Hypericum and hypericin in patients

First author, year	N	Daily dose	Duration	Symptoms
Golsch, 1997 [18]	1	240 mg of extract	3 years	Itchy rash on the face and neck
Bove, 1998 [17]	1	500 mg of powder (equivalent to approx. 100 mg of extract)	4 weeks	Neuropathy of the face and hands
Gulick, 1999 [19]	23	6–12 mg hypericin i.v., corresponding to approx. 10–20 g of *Hypericum* extract	8–24 weeks	Phototoxic skin reactions in 11 out of 23 patients

During a Phase I study carried out in Minnesota with AIDS patients, some 20–40 mg of hypericin were injected intravenously twice a week for up to 24 weeks [19] corresponding to a mean daily dose of 6–12 mg of hypericin. 11 out of 23 patients developed severe phototoxic reactions. As a comparison: the daily recommended dose of the preparation Jarsin® equivalent to 900 mg extract provides approximately 2.8 mg total hypericin. Given a bioavailability of about 20% [15] this corresponds to a systemically available amount of about 0.6 mg total hypericin per day. Therefore it can be estimated that serious phototoxic reactions are likely with about 10–20 times the normal daily dose of 900 mg of extract.

Based on the results of Brockmöller et al. [15], the threshold for a still harmless photosensitisation ought to lie at about 1,800 mg of extract per day. At the present time, this dose should be set as the maximum safely tolerated limit. If overdoses, e.g., ingestion of entire pack contents, are taken with suicidal intent there is a high probability of phototoxic reactions on exposure to sunlight. In such cases, the patients must be protected from daylight and other UV radiation immediately, and due to the long half life of the hypericins (2–3 days), this protection should continue for 1–2 weeks.

Next to this, co-medication with other drugs that possess photosensitising potency may be ill advised. A 44 year-old depressed patient, treated with dothiepin, suffered from a burning and scaling erythema in light exposed but also light unexposed skin, which started after 4 days of concomitant treatment with St. John's Wort extract [20]. Severe erythematobullos dermatosis in a 52 year-old woman may have been the result of concomitant use of oral and topically applied St. John's Wort oil in combination with chloroquine and corticosteroid cream [21]. Furthermore, delta-amino-laevulinic acid, used as a photodiagnostic agent for cancer disease, led to a burning erythematous rash and swelling in a 47 year-old female patient, receiving *Hypericum* extract [13].

Further pharmacodynamic interactions in patients

Six patients suffered from symptoms of serotonin syndrome when St. John's Wort extracts were combined with various drugs affecting central serotonergic activity [22, 23]. The clinical picture, which is supposedly due to central and peripheral serotonergic hyperstimulation, was one of increased somatic disorders, motor restlessness with tremor and gastrointestinal (GI) symptoms. The symptoms stopped when one of the combination partners was discontinued. Serotonin syndrome had also been supposed in three patients who took St. John's Wort without further antidepressive co-medication [24–26]. This supports the notion that at least part of the antidepressive action of St. John's Wort is related to serotonin reuptake inhibition or regulation of serotonin receptors [27, 28]. Therefore it seems reasonable to avoid a combination of St. John's Wort extracts with other antidepressants in order to reduce the risk for this adverse event.

Hypomania or mania had been associated with the intake of St. John's Wort in 17 patients [25, 29–34]. In only two of them, co-medication including antidepressants, was applied [33, 34]. Furthermore, agitation was reported in seven patients after St. John's Wort was taken with unspecified co-medication [25]. In contrast to the latter cases, symptoms of sedation and lethargy had been observed in a male patient, 10 days after he had switched medication from paroxetine to St. John's Wort [35]. St. John's Wort intake had also been held responsible for delayed emergence after surgery in a patient receiving various anaesthetics [36]. Modulation of GABAergic neurotransmission by *Hypericum* has been speculated as reason for this interaction. However, a placebo controlled double-blind crossover study did not reveal an interaction of *Hypericum* extract with ethanol, a known $GABA_A$ receptor activator [37]. No differences in cognitive capacities, measured by use of psychometric tests, were observed in 32 healthy volunteers either receiving a 7 day multiple dose treatment with *Hypericum* or placebo followed by a treatment with alcohol.

An influence of St. John's Wort on the cardiovascular system had been suggested in several case reports. Cardiovascular collapse and hypotension during surgery were ascribed to the intake of St. John's Wort in combination with anaesthetics in a female patient [38]. Hypertension, on the other hand, was observed when *Hypericum* was given alone [39] or in combination with hydrochlorothiazide, oral contraceptives, kava and valerian. Furthermore, a case of hypertensive crisis and one patient with tachycardia and hypertension in relation to an episode of anxiety had been reported [40, 41]. An influence on thyroid hormones by St. John's Wort intake was suggested [42, 43], which might explain some of the symptoms observed in the case reports. However, two prospective clinical studies could not reveal significant effects of St. John's Wort extracts on electrocardiogram in healthy volunteers as well as in depressed patients while treatment with imipramine showed significant increases in first degree AV-block and repolarisation disorders. With regard to heart functions, *Hypericum* extract should therefore be regarded as safer than the tricyclic antidepressants, especially in patients with pre-existing conduction disorders [44, 45].

Pharmacokinetic interactions with other drugs

Recently five reviews [46–50] reported further cases of interactions with coumarin-type anticoagulants, cyclosporin, HIV protease inhibitors as well as suspected cases with theophylline and contraceptives. It is now well established that a multiple dose treatment with St. John's Wort extracts may lead to a decrease in the plasma concentration and efficacy of a variety of co-medicated drugs. *Vice versa*, withdrawal of the herb may generate an increase in concentration and toxicity of medications given concurrently. Naturally, this kind of herbal interaction had attracted particular attention with drugs that are subject to close monitoring because of narrow therapeutic ranges. To evaluate the general clinical relevance of suspected interactions it is crucial to assess each drug class separately.

Cyclosporin and tacrolismus

70 patients have been reported to date, in whom co-medication with St. John's Wort extracts resulted in a significant decrease in blood concentrations of the immunosuppressant cyclosporin. This interaction had also been held responsible for acute transplant rejections in six patients [51–54]. In support of these data, two clinical studies provided evidence of the strong influence of St. John's Wort on cyclosporin concentrations. Multiple dose treatment with *Hypericum* extract led to a significant decrease of single dose kinetics of cyclosporin in healthy volunteers [55]. Similarly, a 14 day multiple dose treatment with St. John's Wort resulted in a marked decrease in the steady state concentrations of cyclosporin in renal transplant patients [56]. Due to similarities in metabolism with cyclosporin, an influence of St. John's Wort had also been anticipated for the immunosuppressant tacrolimus [57, 58]. Indeed, a clinical study in renal transplant patients revealed a comparably strong decrease in tacrolimus blood concentrations after 14 days of treatment with St. John's Wort extract [59]. In contrast, plasma concentration of mycophenolic acid, the active metabolite of the immunosuppressant mycophenolate mofetil, remained stable during treatment with the herb [59].

Warfarin and phenprocoumon

An increase in prothrombin time or a corresponding decrease in the international normalised ratio (INR) had been observed in 29 patients treated with coumarin derivates such as warfarin or phenprocoumon. An increase in INR, as observed in two patients, may possibly be related to withdrawal of the herb during warfarin medication [57]. A clinical study could demonstrate a reduction of the non-protein bound, pharmacologically active concentration of phenprocoumon after 10 days of treatment with St. John's Wort extract in 10 healthy volunteers [60]. However, no report on thrombotic events or bleeding episodes due to co-medication with St. John's Wort extract has been communicated.

HIV-protease inhibitors

A prospective clinical trial found a 57% decrease in steady state concentrations of the HIV-protease inhibitor indinavir in eight healthy volunteers following concomitant treatment with St. John's Wort extract for 14 days [61]. Possibly as a result of decreased plasma concentrations due to St. John's Wort, an increase in HIV viral load had been observed in a patient receiving HIV-triple therapy consisting of indinavir, lamivudine, and stavudine [57]. Furthermore, five patients had been identified during routine drug monitoring, who had taken St. John's Wort in combination with the non-nucleotide reverse transcriptase inhibitor nevirapine [62]. As a result neurapine clearance increased by 35%. These results indicate that St. John's Wort must not be used in HIV infected patients, because of its potential to induce treatment failure and drug resistance.

Oral contraceptives

Second most frequent interactions of St. John's Wort had been communicated with oral contraceptives. Women experienced breakthrough bleeding, spotting or changed menstrual patterns with concurrent use of St. John's Wort extracts suggesting a decrease in the efficacy of oral contraceptive medication. So far, unwanted pregnancies have been related to St. John's Wort co-medication in 11 women [63–65]. However the results of clinical studies investigating the effect of treatment with *Hypericum* extracts on combination oral contraceptives, are contradictory. The preliminary evaluation of ethinylestradiol concentrations in 10 women did not reveal significant alterations during concomitant treatment with *Hypericum* [66]. In two further open studies groups of 12 and 17 young women took a combination of either 20 μg ethinylestradiol and 150 μg desogestrel or 35 μg ethinylestradiol and 1 mg norethinodron over three menstrual cycles. The first menstrual cycle was regarded as control, in the second and third cycles either 600–900 mg or 900 mg *Hypericum* dry extract was taken daily. Measurement of endogenous hormone levels (oestradiol and progesterone; primary endpoints) showed that none of the 29 subjects ovulated in the 3-month study period. There was a significant decrease of about 40% for the area under the curve (AUC) of desogestrel (as the metabolite 3-ketodesogestrel), blood levels of ethinylestradiol and norethinodron were hardly changed. The secondary parameters oral clearance and serum half life showed a moderate increase in the excretion of ethinylestradiol, desogestrel and norethinodron after treatment with *Hypericum* extract. In addition, an increase in breakthrough bleeding (2 to 3 times more frequently than in the control cycle) was reported in both studies. The authors thus suggest that non-compliance as a consequence of increased bleeding carries a risk for undesired pregnancies [67, 68]. However this hypothesis is questionable since 20–60% of women using such contraceptives complain of interim bleeding even when no enzyme-inducing co-medication is being used [69]. Possibly these effects – which were purely subjectively reported by the women – were due to a nocebo-effect (subjects were informed of the possibility at the start of the study, study design was not blind). Similar observations are frequently made in healthy volunteer studies and are well known in the literature [70, 71].

A quite different assessment as an indicator of possible interactions is required from the eight spontaneous reported cases of breakthrough bleeding according to Table 2. Almost three-quarters of all users of *Hypericum* preparations are women. This means that of the roughly 9 million people who took Jarsin® between 1991 and 1999 according to Table 2, at least 4 million were women of child-bearing age. That equates to one report of breakthrough bleeding per half a million treated women in this age group. This incidence is several powers of 10 lower than the spontaneous rate to be expected during the ingestion of low dose oestrogen preparations [69]. Therefore, it must be suspected that some of the reports of breakthrough bleedings and unwanted pregnancies may have been triggered by adverse publicity to St. John's Wort during the last years. On the basis of current knowledge, reduced efficacy of oral contraceptives due to treatment with St. John's Wort should be taken into account, though further studies are needed to validate this herb drug interaction.

Digoxin

The first prospective interaction study of St. John's Wort extract had been performed with the cardiac glycoside digoxin [72]. This parallel study revealed a 25% decrease in steady state concentrations of digoxin in the group of healthy volunteers who received *Hypericum* extract for 10 days compared to placebo. Likewise effects on digoxin pharmacokinetics had been confirmed in two further clinical studies [73, 74]. However, treatments with various doses of the crude drug, *Hypericum* oil, *Hypericum* tea, or a different alcoholic extract of St. John's Wort either did not or only marginally affect digoxin plasma concentrations [74]. These study results indicated that variation in the extent of interaction possibly depends on differences in the composition of St. John's Wort extracts. So far, an interaction with digoxin had not been observed in patients.

Other drugs

A severe interaction of St. John's Wort with medication used in cancer chemotherapy had also been demonstrated, recently. In an open label, crossover study with five cancer patient, co-administration of St. John's Wort extract for 18 days led to a 42% decrease in the systemic exposure of SN-38, the active metabolite of irinotecan [75]. Effects of St. John's Wort medication persisted for 3 weeks after medication with the herbal medicinal was stopped.

A 14 day co-treatment with St. John's Wort extract gave rise to a decrease of steady state plasma concentrations of the tricyclic antidepressant amitriptyline in 12 depressed patients [76]. Concentrations of amitriptyline's primary and active metabolite notriptyline markedly decreased by herbal co-medication. In contrast, administration of St. John's Wort did not affect steady state pharmacokinetics of carbamazepine in eight healthy volunteers [77].

Carbamazepine, which is used in epilepsy and bipolar disorders, is known to induce its own metabolism. Therefore, it had been suggested that the interaction potential of St. John's Wort extract may not be powerful enough to further increase the clearance of carbamazepine after multiple dosing.

St. John's Wort had also been studied to evaluate its influence on two HMG-CoA-reductase inhibitors, pravastatin and simvastatin [78]. In two double-blind, placebo controlled, studies, groups of eight healthy volunteers each were exposed to single doses of simvastatin or pravastatin before and after a 14 day treatment with St. John's Wort. *Hypericum* significantly lowered concentration of simvastatin hydroxy acid, the active metabolite of simvastatin, whereas pravastatin had not been influenced. Reduction of plasma concentrations of theophylline due to treatment with St. John's Wort had been described in two patients [57, 79]. However, these data have not yet been confirmed by a clinical study.

Mechanism and clinical relevance of pharmacokinetic interactions

The majority of drugs and other xenobiotics, like food constituents, exhibit lipophilic characteristics to promote their passage through biological membranes and subsequent access to their site of action. Transformation into more hydrophilic metabolites is therefore essential for the elimination from the body and termination of their biological activity. Such biotransformation relieves the burden of foreign chemicals and is critical for the survival of the organism. Studies of the genes that encode the enzymes of biotransformation have led to the view that they evolved more than one billion years ago as a mechanism for removal of natural constituents of food, such as flavones, terpenes, steroids, and alkaloids. This has led to the remarkable overlap of substrate specificities of those enzymes. The cytochrome P450 monooxygenase system is the most important biotransformation system, responsible for the inactivation of a wide range of exogenous and endogenous (e.g., steroids) compounds. It is mainly located in the liver, in enterocytes and the kidney, and isoforms CYP3A, CYP2D, and CYP2C account for the metabolism of 50%, 25%, and 20%, respectively, of the currently known drugs [80, 81]. More recently, transporter proteins located in cell membranes have gained interest in relation to drug distribution and elimination. An important representative from this group seems to be the efflux transporter P-glycoprotein (P-gp), which is encoded by the multidrug resistance-gene 1 (MDR-1). P-gp is present in liver, kidney, and at the blood–brain barrier and at several other sites, and constitutes, along with intestinal CYP 450 metabolism, an important part of the biochemical barrier of the intestinal mucosa [82].

The interactions associated with St. John's Wort have been diligently recorded and researched in recent years and described in more than 50 publications. The reader may consult current reviews for additional details [12, 46–50]. Results of these studies suggest that the mechanism of pharmacokinetic interactions with St. John's Wort extract involves the drug-metabolising enzyme cytochrome P-450 (CYP) isoenzymes (mainly CYP3A4), and the transport protein P-glycoprotein. To be fair, it should be mentioned that the activation or inhibition of the CYP enzymes is a very non-specific reaction by the body that is also changed by a great many other herbs, spices, and food products [83], among them grapefruit [84], garlic [85], black tees [86], honey [87, 88] and red wine [89]. For the clinician, the key question is the significance of pharmacokinetic results obtained in the laboratory, especially those measured in healthy subjects. Changes in plasma concentrations, like those shown with St. John's Wort for phenprocoumon, digoxin or amitriptyline, are not unusual in everyday practice. Corresponding interactions with the ingestion of food have been described for more than 50 drugs [90]. No fewer than 152 herbal drugs are suspected of interacting with synthetic drugs. However, less than one tenth of the theoretical suspected cases were associated with clinical references [8, 46, 47].

Based on current clinical and experimental data, co-medication with St. John's Wort extract should be avoided in patients who are taking cyclosporin- or tacrolimus-type immunosuppressant drugs with a narrow therapeutic range, coumarin-type anticoagulants, or virostatic drugs in the form of protease inhibitors and/or non-competitive reverse transcriptase inhibitors as well as in patients receiving cancer chemotherapy with irinotecan. The possibility of an interaction with oral contraceptives has to be kept in mind, but needs further research. St. John's Wort extracts, used in the treatment of mild-to-moderate depression, should be given in monotherapy, because of their potential to induce serotonin syndrome when combined with standard antidepressants. The possibility of photosensitivity reactions due to St. John's Wort should also be taken into account, especially when therapy is combined with laser treatment or with drugs that are known to increase photosensitivity.

References

1 Harrer G, Payk TR, Schulz V (Hrsg.) (1993) *Hypericum* als pflanzliches Antidepressivum. *Nervenheilkunde* 12: 268–366

2 Jenike MA (ed.) (1994) Hypericum: A noval antidepressant. *J Geriatr Psychiatry Neurol* 7 (Suppl 1): 1–68

3 Müller WE, Kasper S (eds) (1997) *Hypericum* extract (LI 160) as a herbal antidepressant. *Pharmacopsychiatry* 30 (Suppl II): 71–134

4 Müller WE, Chatterjee SS (eds) (1998) Hyperforin and the antidepressant activity of St. John's Wort. *Pharmacopsychiatry* 31 (Suppl I): 1–60

5 Knüppel L, Linde K (2004) Adverse effects of St John's Wort – An overview. *J Clin Psychiatry* 65: 1470–1479

6 Volz HP, Zeller K (2000) Hochdosierter Johanniskrautextrakt als Einmalgabe. Ergebnisse einer Anwendungsbeobachtung an 11296 Patienten. *Nervenheilkunde* 19: 566–569

7 Woelk H, Burkhard G, Grünwald J (1994) Benefits and risks of the *Hypericum* extract LI 160: Drug monitoring study with 3250 patients. *J Geriatr Psychiatry Neurol* 7 (Suppl 1): 34–38

8 Schulz V (2001) Incidence and clinical relevance of the interactions and side effects of *Hypericum* preparations. *Phytomedicine* 8: 152–160

9 Araya OS, Ford EJH (1981) An investigation of the type of photosensitization caused by the inges-

tion of St. John's Wort (*Hypericum perforatum*) by calves. *J Comp Pathol* 91: 135–141
10 Giese AC (1980) Hypericism. *Photochem Photobiol Rev* 39: 229–255
11 Fugh-Berman A, Ernst E (2001) Herb-drug interactions: review and assessment of report reliability. *Br J Clin Pharmacol* 52: 587–595
12 Mannel M (2004) St. John's Wort drug interactions – review of mechanisms and clinical implications. *Drug Saf* 27: 773–797
13 Ladner DP, Klein SD, Steiner RA, Walt H (2001) Synergistic toxicity of delta-aminolaevulinic acid-induced protoporphyrin IX used for photodiagnosis and *Hypericum* extract, a herbal antidepressant. *Br J Dermatol* 144: 916–918
14 NN (1984) Monographie Hyperici herba (Johanniskraut). *Bundesanzeiger* Nr. 228, 05.12.1984
15 Brockmöller J, Reum T, Bauer S, Kerb R, Hübner WD, Roots I (1997) Hypericin and pseudohypericin: Pharmacokinetics and effects on photosensitivity in humans. *Pharmacopsychiatry* 30 (Suppl): 94–101
16 Schempp CM, Winghofer B, Müller K, Schulte-Mönting J, Mannel M, Schöpf E, Simon J (2003) Effect of oral administration of *Hypericum perforatum* extract (St. John's Wort) on skin erythema and pigmentation induced by UVB, UVA, visible light and solar simulated radiation. *Phytotherapy Res* 17: 141–146
17 Bove GM (1998) Acute neuropathy after exposure to sun in a patient treated with St John's Wort. *Lancet* 352: 1121–1122
18 Golsch S, Vocks E, Rakoski J, Brockow K, Ring J (1997) Reversible Erhöhung der Photosensitivität im UV-B-Bereich durch Johanniskraut-Präparate. *Der Hautarzt* 48: 249–252
19 Gulick RM, McAuliffe V, Holden-Wiltse J, Crumpacker C, Liebes L, Stein DS, Meehan P, Hussey S, Forcht J, Valentine FT (1999) Phase I studies of hypericin, the active compound in St. John's Wort, as an antiretroviral agent in HIV-infected adults. *Ann Intern Med* 130: 510–514
20 Holme SA, Roberts DL (2000) Erythroderma associated with St John's Wort. *Br J Dermatol* 143: 1127–1128
21 Lane-Brown MM (2000) Photosensitivity associated with herbal preparations of St John's Wort (*Hypericum perforatum*). *Med J Aust* 172: 302
22 Gordon J (1998) SSRIs and St John's Wort: possible toxicity? *Am Fam Physician* 62: 31
23 Lantz MS, Buchalter E, Giambanco V (1999) St John's Wort and antidepressant drug interactions in the elderly. *J Geriatr Psychiatry Neurol* 12: 7–10
24 Demott K (1998) St. John's Wort tied to serotonine syndrome. *Clin Psychiatry News* 26: 28
25 Walter G, Rey JM, Harding A (2000) Psychiatrists' experience and views regarding St John's Wort and 'alternative' treatments. *Aust N Z J Psychiatry* 34: 992–996
26 Parker V, Wong AH, Boon HS, Seeman MV (2001) Adverse reactions to St John's Wort. *Can J Psychiatry* 46: 77–79
27 Greeson JM, Sanford B, Monti DA (2001) St. John's Wort (*Hypericum perforatum*): a review of the current pharmacological, toxicological, and clinical literature. *Psychopharmacology* 153: 402–414
28 Müller WE (2003) Current St. John's Wort research from mode of action to clinical efficacy. *Pharmacol Res* 47: 101–109
29 O'Breasail AM, Argouarch S (1998) Hypomania and St John's Wort. *Can J Psychiatry* 43: 746–747
30 Schneck C (1998) St. John's Wort and hypomania. *J Clin Psychiatry* 59: 689
31 Nierenberg AA, Burt T, Matthews J, Weiss AP (1999) Mania associated with St. John's Wort. *Biol Psychiatry* 46: 1707–1708
32 Moses EL, Mallinger AG (2000) St. John's Wort: three cases of possible mania induction. *J Clin Psychopharmacol* 20: 115–117
33 Barbenel DM, Yusufi B, O'Shea D, Bench CJ (2000) Mania in a patient receiving testosterone replacement postorchidectomy taking St John's Wort and sertraline. *J Psychopharmacol* 14: 84–86
34 Spinella M, Eaton LA (2002) Hypomania induced by herbal and pharmaceutical psychotropic medicines following mild traumatic brain injury. *Brain Inj* 16: 359–367
35 Gordon JB (1998) SSRIs and St. John's Wort: possible toxicity? *Am Fam Physician* 57: 950
36 Crowe S, McKeating K (2002) Delayed emergence and St. John's Wort. *Anesthesiology* 96: 1025–1027
37 Schmidt U, Harrer G, Kuhn U, Berger-Deinert W, Luther D (1993) Wechselwirkungen von *Hypericum*-Extrakt mit Alkohol. *Nervenheilkunde* 12: 314–319

38 Irefin S, Sprung J (2000) A possible cause of cardiovascular collapse during anesthesia: long- term use of St. John's Wort. *J Clin Anesth* 12: 498–499
39 Zullino D, Borgeat F (2003) Hypertension induced by St. John's Wort. *Pharmacopsychiatry* 36: 32
40 Patel S, Robinson R, Burk M (2002) Hypertensive crisis associated with St. John's Wort. *Am J Med* 112: 507–508
41 Brown TM (2000) Acute St. John's Wort toxicity. *Am J Emerg Med* 18: 231–232
42 Ferko N, Levine MA (2001) Evaluation of the association between St. John's Wort and elevated thyroid-stimulating hormone. *Pharmacotherapy* 21: 1574–1578
43 Hauben M (2002) The association of St. John's Wort with elevated thyroid-stimulating hormone. *Pharmacotherapy* 22: 673–675
44 Siepmann M, Krause S, Joraschky P, Muck-Weymann M, Kirch W (2002) The effects of St John's Wort extract on heart rate variability, cognitive function and quantitative EEG: a comparison with amitriptyline and placebo in healthy men. *Br J Clin Pharmacol* 54: 277–282
45 Czekalla J, Gastpar M, Hubner WD, Jager D (1997) The effect of *Hypericum* extract on cardiac conduction as seen in the electrocardiogram compared to that of imipramine. *Pharmacopsychiatry* 30 (Suppl 2): 86–88
46 Ernst E (2000) Possible interactions between synthetic and herbal medicinal products. Part 1: a systematic review of the indirect evidence. *Perfusion* 13: 4–15
47 Ernst E (2000) Interactions between synthetic and herbal medicinal products. Part 2: a systematic review of the direct evidence. *Perfusion* 13: 60–70
48 Fugh-Berman A (2000) Herb drug interactions. *Lancet* 355: 134–138
49 Johne A, Schmider J, Maurer A, Brockmöller J, Mai I, Donath F, Roots I (2000) Arzneimittel-Interaktionen mit Johanniskraut. In: N Rietbrock (ed.): *Phytopharmaka in Forschung und klinischer Anwendung VI.* Steinkopff-Verlag, Darmstadt, 55–66
50 Henderson L, Yoe QY, Bergquist C, Gerden B, Arlett P (2002) St John's Wort (*Hypericum perforatum*): drug interactions and clinical outcomes. *Br J Clin Pharmacol* 54: 349–356
51 Ruschitzka F, Meier PJ, Turina M, Luscher TF, Noll G (2000) Acute heart transplant rejection due to Saint John's Wort. *Lancet* 355: 548–549
52 Breidenbach T, Hoffmann MW, Becker T, Schlitt H, Klempnauer J (2000) Drug interaction of St. John's Wort with cyclosporin. *Lancet* 355: 1912
53 Karliova M, Treichel U, Malago M, Frilling A, Gerken G, Broelsch CE (2000) Interaction of *Hypericum perforatum* (St. John's Wort) with cyclosporin A metabolism in a patient after liver transplantation. *J Hepatol* 33: 853–855
54 Barone GW, Gurley BJ, Ketel BL, Lightfoot ML, Abul-Ezz SR (2000) Drug interaction between St. John's Wort and cyclosporine. *Ann Pharmacother* 34: 1013–1016
55 Dresser GK, Schwarz UI, Wilkinson GR, Kim RB (2003) Coordinate induction of both cytochrome P4503A and MDR1 by St John's Wort in healthy subjects. *Clin Pharmacol Ther* 73: 41–50
56 Bauer S, Stormer E, Johne A, Kruger H, Budde K, Neumayer HH, Roots I, Mai I (2003) Alterations in cyclosporin A pharmacokinetics and metabolism during treatment with St John's Wort in renal transplant patients. *Br J Clin Pharmacol* 55: 203–211
57 Barnes J, Anderson LA, Phillipson JD (2001) St John's Wort (*Hypericum perforatum* L.): a review of its chemistry, pharmacology and clinical properties. *J Pharm Pharmacol* 53: 583–600
58 Bolley R, Zulke C, Kammerl M, Fischereder M, Kramer BK (2002) Tacrolimus-induced nephrotoxicity unmasked by induction of the CYP3A4 system with St John's Wort. *Transplantation* 73: 1009
59 Mai I, Störmer E, Bauer S, Krüger H, Budde K, Roots I (2003) Impact of St John's Wort treatment on the pharmacokinetics of tacrolimus and mycophenolic acid in renal transplant patients. *Nephrol Dial Transplant* 18: 819–822
60 Maurer A, Johne A, Bauer S, Brockmöller J, Donath F, Roots I, Langheinrich M, Hübner WD (1999) Interaction of St. John's Wort extract with phenprocoumon. *Eur J Clin Pharmacol* 55: A22
61 Piscitelli SC, Burstein AH, Chaitt D, Alfaro RM, Falloon J (2000) Indinavir concentrations and St John's Wort. *Lancet* 355: 547–548
62 de Maat MM, Hoetelmans RM, Math t RA, van Gorp EC, Meenhorst PL, Mulder JW, Beijnen JH (2001) Drug interaction between St John's Wort and nevirapine. *Aids* 15: 420–421
63 Anonymus (2001) Vorsicht: "Pillen"-Versager durch Johanniskraut. *Arznei-Telegramm* 32: 35
64 Schwarz UI, Büschel B, Kirch W (2001) Failure of oral contraceptives because of St. John's Wort.

Eur J Clin Pharmacol 57: A25
65 Henderson L, Yue QY, Bergquist C, Gerden B, Arlett P (2002) St John's Wort (*Hypericum perforatum*): drug interactions and clinical outcomes. *Br J Clin Pharmacol* 54: 349–356
66 Käufeler R, Meier B, Brattström A (2000) Johanniskrautextrakt Ze 117 – Klinische Wirksamkeit und Verträglichkeit. In: N Reitbrock, F Donath, D Loew, I Roots, V Schulz (eds): *Phytopharmaka VI: Forschung und Klinische Anwendung*. Steinkopff, Darmstadt, 84–89
67 Pfrunder A, Schiesser M, Gerber S, Haschke M, Bitzer J, Drewe J (2003) Interaction of St John's Wort with low-dose oral contraceptive therapy: a randomized controlled trial. *Br J Cin Pharmacol* 56: 683–690
68 Hall SD, Wang Z, Huang SM, Hamman MA, Vasavada N, Adigun AQ, Hilligoss JK, Miller M, Gorski JC (2003) The interaction between St John's Wort and oral contraceptive. *Clin Pharmacol Ther* 74: 525–535
69 Rosenberg MJ Waugh MS, Higgins SE (1996) The effect of desogestrel, gestadone, and other factors on spotting and bleeding. *Contraception* 53: 85–90
70 Weihrauch TR, Gauler TC (1999) Placebo – Efficacy and adverse effects in controlled clinical trials. *Arzneim-Forsch/Drug Res* 49: 385–393
71 Pogge RC (1963) The toxic placebo. Part I – Side and toxic effects reported during the administration of placebo medicine. Part II – The case of the partial patient. *Medical Times* 91: 773–781
72 Johne A, Brockmöller J, Bauer S, Maurer A, Langheinrich M, Roots I (1999) Pharmacokinetic interaction of digoxin with an herbal extract from St Johns Wort (*Hypericum perforatum*). *Clin Pharmacol Ther* 66: 338–345
73 Durr D, Stieger B, Kullak-Ublick GA, Rentsch KM, Steinert HC, Meier PJ, Fattinger K (2000) St John's Wort induces intestinal P-glycoprotein/MDR1 and intestinal and hepatic CYP3A4. *Clin Pharmacol Ther* 68: 598–604
74 Mueller SC, Uehleke B, Woehling H, Petzsch M, Majches-Perszynska J, Hehl EM, Sievers H, Frank B, Riethling AK, Drewelow B (2004) Effect of St. John's Wort dose and preparations on the pharmacokinetics of digoxin. *Clin Pharmacol Ther* 75: 546–557
75 Mathijssen RH, Verweij J, de Bruijn P, Loos WJ, Sparreboom A (2002) Effects of St. John's Wort on irinotecan metabolism. *J Natl Cancer Inst* 94: 1247–1249
76 Johne A, Schmider J, Brockmöller J, Stadelmann AM, Störmer E, Bauer S, Scholler G, Langheinrich M, Roots I (2002) Decreased plasma levels of amitriptyline and its metabolites on comedication with an extract from St. John's Wort (*Hypericum perforatum*). *J Clin Psychopharmacol* 22: 46–54
77 Burstein AH, Horton RL, Dunn T, Alfaro RM, Piscitelli SC, Theodore W (2000) Lack of effect of St John's Wort on carbamazepine pharmacokinetics in healthy volunteers. *Clin Pharmacol Ther* 68: 605–612
78 Sugimoto K, Ohmori M, Tsuruoka S, Nishiki K, Kawaguchi A, Harada K, Arakawa M, Sakamoto K, Masada M, Miyamori I, Fujimura A (2001) Different effects of St John's Wort on the pharmacokinetics of simvastatin and pravastatin. *Clin Pharmacol Ther* 70: 518–524
79 Nebel A, Schneider BJ, Baker RK, Kroll DJ (1999) Potential metabolic interaction between St. John's Wort and theophylline. *Ann Pharmacother* 33: 502
80 Wilkinson GR. Pharmacokinetics. In: JG Hardman, LE Limbird (eds): *Goodman and Gilman's the pharmacological basis of therapeutics*. MacGraw-Hill, New York, 3–29
81 Rendic S, Di Carlo FJ (1997) Human cytochrome P450 enzyme: a status report summarizing their reactions, substrates, induction, and inhibitors. *Drug Metab Rev* 29: 413–580
82 Bodo A, Bakos E, Szeri F (2003) The role of multidrug transporters in drug availability, metabolism and toxicity. *Toxicol Lett* 141: 133–143
83 Budzinski JW, Foster BC, Vandenhoek S, Arnason JT (2000) An *in vitro* evaluation of human cytochrome P450 3A4 inhibition by selected commercial herbal extracts and tinctures. *Phytomedicine* 7: 273–282
84 Ameer B, Weintraub RA (1997) Drug interaction with grapefruit juice. *Clin Pharmacokin* 33: 103–121
85 Foster BC, Foster MS, Vandenhoek S, Krantis A, Budzinski JW, Arnason JT, Gallicano KD, Choudri S (2001) An *in vitro* evaluation of human cytochrome P450 3A4 and P-glycoprotein inhibition by garlic. *J Pharm Pharmac Sci* 4: 176–184
86 Foster BC, Vandenhoek S, Hana J, Krantis A, Akhtar MH, Bryan M, Budzinski JW, Ramputh A, Arnason JT (2003) *In vitro* inhibition of human cytochrome P450-mediated metabolism of marker substrates by natural products. *Phytomedicine* 10: 334–342

87 Koumaravelou K, Adithan C, Shashindran CH, Asad M, Abraham BK (2002) Effect of honey on carbamazepine kinetics in rabbits. *Indian J Exp Biol* 40: 560–563
88 Koumaravelou K, Adithan C, Shashindran CH, Asad M, Abraham BK (2002) Influence of honey on orally and intravenously administered diltiazem kinetics in rabbits. *Indian J Exp Biol* 40: 1164–1168
89 Tsunoda SM, Harris RZ, Christians U, Velez RL, Freeman RB, Benet LZ, Warshaw A (2001) Red wine decreases cyclosporine bioavailability. *Clin Pharmacol Ther* 70: 462–467
90 Williams L Hill DP, Davis JA, Lowenthal DT (1996) The influence of food on the absorption and metabolism of drugs: an update. *Eur J Drug Metab Pharmacokinet* 21: 201–211

St. John's Wort and its Active Principles in Depression and Anxiety
Edited by W. E. Müller

Endocrinology of St. John's Wort extract

Michael Franklin

University of Oxford Department of Psychiatry, Warneford Hospital, Oxford OX3 7JX, United Kingdom; and School of Biological and Molecular Sciences, Oxford Brookes University, Oxford OX3 0BP, United Kingdom

Introduction

The endocrine system is one of the two great control systems of the body, the other being the nervous system. These two regulatory systems are responsible for monitoring changes in an animal's internal and external environments and directing the body to make necessary adjustments to its activities so that it adapts itself to these environmental changes. The endocrine system operates through chemical messengers, or hormones, which circulate in the blood to their respective target organs and modify their activity. The term "hormone" was introduced in 1905 by Starling and is derived from the Greek, meaning "to arouse", though it should be said that not all hormonal effects are stimulatory; indeed some are inhibitory. The endocrine glands secrete hormones into the circulation.

The release of these hormones into the circulation by endocrine glands may be influenced by external environmental factors such as stress. The endocrine system responds to recurrent changes, such as meals, and helps the organism to adapt to changing habits. It is also important for controlling development and growth, puberty and sexual maturation. Therefore, the secretion of hormones, apart from maintaining the body's internal environment, can also induce important long-term changes.

For the body of this text we will concentrate on hormones such as prolactin and cortisol released from the pituitary and adrenal glands respectively. Hormones such as prolactin, growth hormone (GH) etc., are glycoprotein hormones secreted from the anterior pituitary and are modulated by releasing or inhibiting hormones from the hypothalamus. For example, hormones such as somatostatin and growth hormone releasing factor (GHRF) inhibit and stimulate the release of GH respectively. Furthermore. monoamine neurotransmitter pathways are involved in the release of these modulating hormones. Monoamines such as serotonin (5-hydroxytryptamine, 5-HT), containing neurones, innervate hypothalamic nuclei containing the cell bodies of the anterior pituitary hormone releasing (and inhibitory) factor neurones. These send projections to the median eminence, the terminals of which abut the capillaries of the hypothalamo–hypophyseal–portal system. Stimulation of these neurones causes secretion of the releasing or inhibitory factors into the hypothalamic-portal system from where they are transported to the anterior pituitary. Activation of a monoamine pathway by a drug with a specific action on that pathway leads to a characteristic pattern of hormone release detectable in venous blood. The magnitude of the hormone response within certain limitations can be used as a pharmacological index of the central action of that drug.

Neuroendocrine challenge tests, which are based on principles mentioned above, will be discussed in the future text [1]. Such tests that utilise specific acting drugs have been used extensively to study 5-HT and other monoamine neurotransmitter functions in various psychiatric disorders. Their rational have been described [2, 3].

Hypericum extract is a pharmacological agent obtained from the plant St John's Wort (*Hypericum perforatum*, HP). The genus *Hypericum* includes more than 400 species of the plant, which grow worldwide. It also exists in a wide range of preparations that differ in the content of known active constituents such as the phloroglucinols (e.g., hyperforin), flavinoids (e.g., hyperosid), napthodianthrones (e.g., hypericin), procyanidines (e.g., procyanidine B2) and xanthones. Therefore, carrying out research studies with such complex agents is obviously very difficult. Likewise comparison of data across studies is also difficult. Hence it is always important to state extract origins and method of preparation when comparing data across the board.

HP is licensed in Germany as an antidepressant. Clinical trials have shown it to be better than placebo and generally equal to antidepressant drugs such as imipramine for the treatment of mild-to-moderate depression [4].

Despite all the clinical trials, a gamete of research studies and also probably due to its overall complex makeup, its clinical efficacy and mode of action is still to be fully elucidated. Likewise its full effects/actions on the endocrine system are unclear.

This review of the hormonal actions of HP will cover both its acute and chronic effects, but will take the view that the chronic effects are the more important. The majority of the studies have been carried out in animals, though some have been done in human volunteers too. Therefore where possible results have been compared and contrasted between the two. There are no recorded studies on the effect of HP in depressed subjects.

Hormones

Although the true mechanism of HP's antidepressant effect has not been fully elucidated, studies have demonstrated that like conventional antidepressant drugs, HP dose-dependently inhibits uptake of 5-HT, noradrenaline (NA) and dopamine (DA) in synaptosomal preparations [5]. Previously, the antidepressant action of HP extracts had been mainly attributed to hypericin on whose content the majority of commercial products have been standardised.

However, recently some emphasis has been placed on another constituent, hyperforin. This is because hypericin demonstrates no monoamine reuptake inhibiting properties whereas hyperforin showed potent reuptake inhibiting effects comparable to conventional antidepressant drugs, although it was not specific to any one monoamine (see Tab. 1). This data therefore suggested that hyperforin was the single most active constituent of HP extracts in biochemical models of antidepressant action. Some extracts, such as LI 160, contain higher amounts of hyperforin than others. This is dependent on the mode of extraction, the part of the plant extracted and the origins of the plant. The neuroendocrine paradigm has been used effectively to establish a measure of the central monoamine activity of psychotropic drugs.

Table 1. Inhibition of the synaptosomal uptake of monoamines by two major constituents of HP extracts, hypericin and hyperforin, and two conventional antidepressant drugs, imipramine and nomifensine. Adapted from [5].

	IC_{50} (nmol/ml)			
Uptake system	Hyperforin	Hypericin	Imipramine	Nomifensine
5-HT	205 ± 45	>10,000	207 ± 10	–
NA	80 ± 20	>10,000	21 ± 17	1.9 ± 1.2
DA	102 ± 17	>10,000	–	37 ± 2.5

Table 2 shows the effect of some antidepressant drugs and other psychoactive agents on hormone release. Similarly this has been used to study HP extract actions on central monoamine function as indexed by their effects on various hormonal outputs.

Cortisol and related hormones

Corticotrophin (ACTH) secreted by the anterior pituitary controls the synthesis and release of steroid hormones such as cortisol/corticosterone from the adrenal cortex. In turn, ACTH is under the control of corticotrophin-releasing factor (CRF) released from the hypothalamus. The adrenal cortex secretes the glucocorticosteroids (GCs) and the mineralocorticoids (MCs). The GCs greatly affect carbohydrate and protein metabolism, whereas the MRs (for example, aldosterone) affect water and electrolyte balance. Cortisol accounts for most of the GCs activity of the adrenal hormones in man and primates although in some species (e.g., rat) corticosterone is the predominant GC. GC release is controlled by negative feedback loops, i.e., cortisol can reduce its own secre-

Table 2. Influence of antidepressants on monoamine reuptake and influence of antidepressants and other psychotropic drugs on growth hormone (GH), prolactin (PRL) and cortisol COR secretion in man. Adapted from [47].

Reuptake inhibition NA	5-HT	DA	GH	PRL	CORT
***	*	Desipramine	***	**	***
***		R-(+)-Oxaprotiline	***	0	
		L-(+)-Oxaprotiline	0	0	0
***		Reboxetine	***	**	***
		Nomifensine	**	?	
	***	Clomipramine	**	***	***
		Indalpine	0	***	***
*	***	Venlafaxine	*	***	***
A2-, 5-HT2- and 5-HT3 blocker		Mirtazepine	0	?	??
MAO inhibitor		Moclobemide	*	***	*
DA blocker		Haloperidol	0	****	0
		Sulpiride	0	****	0
GABA agonist		Diazepam	*	0	(?)
		Metaclazepam	*		

tion rate. CRF is generally under NA and 5-HT release control. Cortisol release is increased under conditions such as stress.

A pilot study from my own laboratory in six healthy male volunteers showed that the HP extract LI 160 (Jarsin, 2,700 mg) when administered as a single dose significantly increased salivary concentrations of cortisol (see Tab. 3) compared to placebo [6].

In another study, the HP extract WS 5570 was administered to healthy male volunteers at two doses 300 mg and 600 mg *versus* placebo on three different days [7]. Serum cortisol concentrations were significantly increased after the higher dose, while after the lower dose they were similar to placebo (see Fig. 1). The authors suggested that since both NA reuptake inhibitors such as desipramine and the 5-HT reuptake inhibitors such as clomipramine are known to increase plasma cortisol acutely [8], it is therefore possible that the cortisol stimulation caused by the higher dose of WS 5570 HP extract may also be influencing the same monoamines, 5-HT and NA. In a more recent study, and as yet unpublished, the same authors were unable to confirm their previous findings on WS5570's effects on cortisol and ACTH excretion [9]. In this new

Table 3. The effect of LI 160 extract (Jarsin) treatment on salivary cortisol in six healthy male volunteer subjects.

Time (h)	12.00	13.00	14.00	15.00	16.00	17.00	18.00
Jarsin	5.3	10.3*	11.0*	7.1	4.5	4.0	2.7
± sem	0.8	2.1	2.2	1.5	0.5	0.4	0.5
Placebo	4.7	4.3	4.5	4.2	2.7	3.0	2.9
± sem	0.4	0.8	1.1	1.0	0.2	0.6	0.3

* $P = 0.04$

Results are presented as mean ± sem µg/ml. Analysis of the salivary cortisol data by repeated measures ANOVA showed that there were significant effects of drug ($F = 22.5$; $df = 5,1$; $p = 0.005$) and time ($F = 3.3$; $df = 30.6$; $p < 0.01$). Adapted from [6].

study they examined the effect of various doses (600, 900 and 1,200 mg) *versus* placebo on serum concentrations of cortisol, ACTH, human growth hormone (HGH) and prolactin. Analysis of the data revealed a significant stimulatory effect on plasma ACTH; however serum cortisol was not affected. Why there is a discrepancy between the two studies is not clear. However, as mentioned previously in this text there is great variation between the active contents of these herbal extracts, and indeed between batches of the same extract and hence it is therefore possible that this may be one reason for the contrasting differences between these studies from the same laboratory.

In a further study of our own, a small subgroup (30%) of normal healthy volunteer subjects who received a single high dose (2,700 mg) treatment of the HP extract LI 160 (Jarsin), showed significant increases in plasma cortisol *ver-*

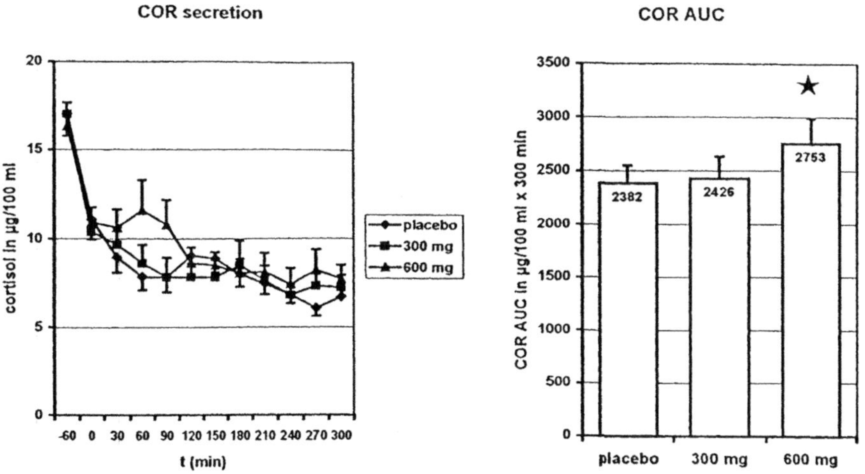

Figure 1. Cortisol (µg/100 ml) after administration of placebo, WS 5570 300 and 600 mg; mean value graphs and AUC graphs; S.E. indicated; * $p < 0.05$ (*versus* placebo). Taken from [7].

sus placebo treatment [10]. However, when all subjects from the study were considered, plasma cortisol was non-significantly increased *versus* placebo, thus suggesting overall that this was not a robust response. In this same study, plasma concentration kinetics of hyperforin, but not hypericin, appeared to synchronise with the timing of the cortisol responses, although no significant correlations were found. The findings of Müller et al. [11], who identified hyperforin and not hypericin as the non-specific reuptake inhibitor of HP extracts, would seem to support the notion that hyperforin may be the most likely constituent of the HP extracts used in these studies causing the plasma cortisol stimulation.

In direct contrast to the findings in human volunteer studies previously discussed, single dose (intraperitoneal, *ip*) studies in the rat carried out in our laboratory showed that the HP extract LI 160 and it's constituents, hyperforin and hypericin, all consistently increased serum corticosterone [12]. Indeed, doses of LI 160 as low as 5 mg caused significant increases in serum corticosterone (see Fig. 2). The HP extract LI 160, hyperforin and hypericin were found to increase serum corticosterone to the same degree at equivalent doses when compared to their own control (i.e., vehicle alone), although numerically hypericin was only 50% as potent as LI 160 and hyperforin.

The reason for the inconsistency between the studies in man and those in the rat is not clear. However, it is possible that the differing routes of administration in these studies may be one confounding factor or it may more simply be due to a species difference.

Studies show that the increase in serum corticosterone produced by the HP extract LI 160 in the rat, could be inhibited by the 5-HT_2-receptor antagonist, ketanserin but not by the 5-HT_{1A}-receptor antagonist, WAY-100635 [12] (see Fig. 3). It is unlikely that LI 160 was acting directly on 5-HT_2 receptors as it

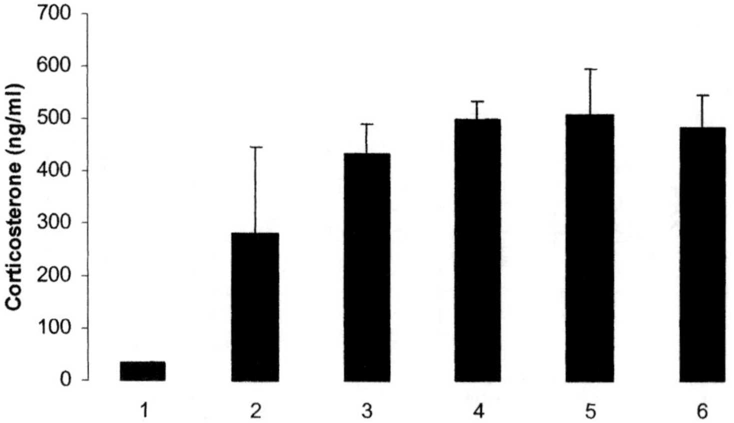

Figure 2. Serum corticosterone responses to various doses of the HP extract, Li 160 in male rats (n = 7 per dose). Results are expressed as mean ± sd [previously unpublished data]. NB. 1 = placebo; 2 = 5 mg/kg; 3 = 10 mg/kg; 4 = 20 mg/kg; 5 = 5 mg/kg; 6 = 100 mg/kg, respectively.

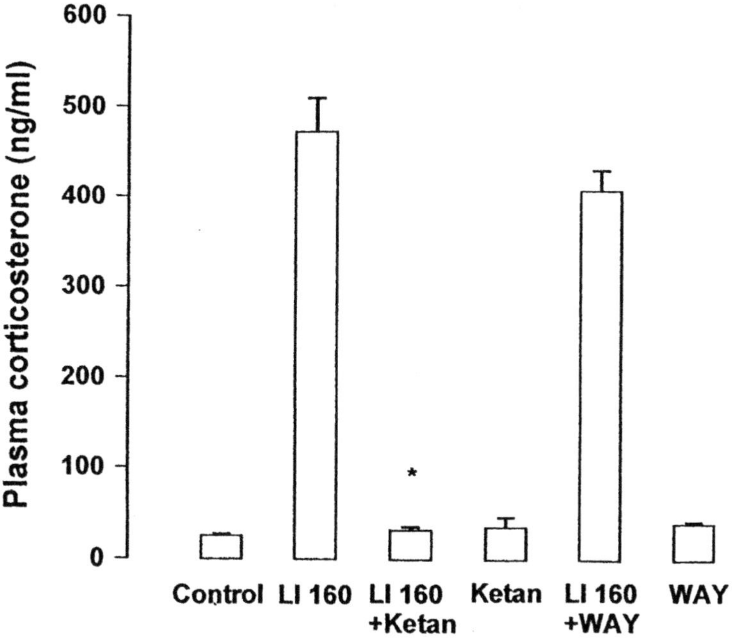

Figure 3. Effect of the 5-HT_{1A} and 5-HT_2 antagonists WAY-100635 (0.5 mg/kg, *ip*) and keteanserin (2 mg/kg, *ip*) respectively on plasma corticosterone responses to acute single dose treatment with LI 160 (200 mg/kg, *ip*). Both antagonists were administered 30 min prior to LI 160 treatment in each instance and the animals being finally sacrificed 2 h later. The control treatment was saline alone. Each bar represents the mean ± sem from five animals. *p = 0.003. Taken from [23].

did not produce any typical behavioural syndrome when given to these animals. It is therefore likely that the HP extract LI 160 produces an indirect activation of post-synaptic 5-HT_2 receptors possibly through inhibition of 5-HT reuptake blockade [5].

Reports of increases in brain monoamine content lend support to the findings above. Both *in vivo* and *in vitro* studies have shown that extracts of HP increase the release of catecholamines into the cortex, diencephalon, brain stem [13], locus coeruleus [14], striatum and nucleus accumbens [15]. More recently a microdialysis study in rat brain showed that the HP extract LI 160, significantly increased both 5-HT and DA, but did not affect NA release [16].

Neuroendocrine studies also suggest that HP extracts exert their antidepressant actions through 5-HT_2 receptors. Serum corticosterone and prolactin responses to the selective 5-HT_{2A} agonist, DOI (2,5-dimethoxy-4-iodophenyl-2-aminopropane) were reduced (see Figs 4 and 5) following 2 weeks of treatment with the HP extract LI 160 in rats [17]. It was suggested that this may be the result of decreased 5-HT_{2A} receptor expression, signal transduction or intracellular messengers. However, since DOI does show some affinity for the 5-HT_{2C} receptor, a role for this receptor cannot be ruled out [18].

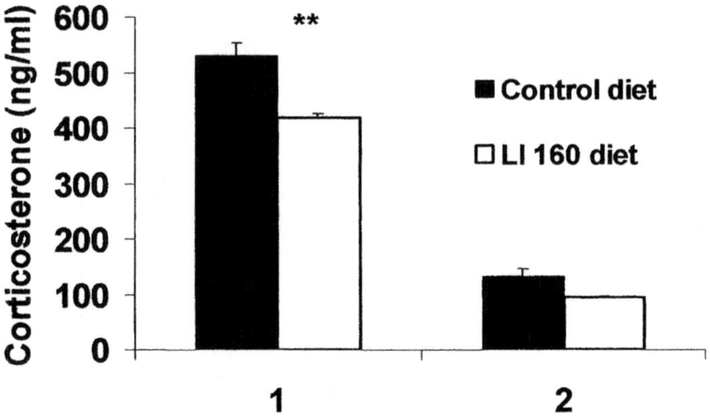

Figure 4. Effect of a diet containing LI 160 (3 g/kg) *versus* control diet on the serum corticosterone responses to the 5-HT_{2A} agonist, DOI (dose = 2 mg/kg *ip*) and saline (*ip*) in male rats (n = 7 per group). Results are given as mean ± sem. ** $p < 0.002$. Taken from [17].

There may be a plausible alternative explanation for reduced neuroendocrine responses to DOI, at least in the case of corticosterone. Firstly, a recent study showed that LI 160 reduces corticosterone-releasing hormone (CRH) mRNA in the hypothalamic paraventricular nucleus after 8 weeks of treatment by gavage [19]. It is difficult to directly compare this study with that described since both the route of administration and the doses used were different. It is possible that the reduced corticosterone responses to DOI found in this study might be due to a reduction in CRH mRNA gene expression and/or the induction of the multi-drug transporter glycoprotein (Pgp) [19, 20]. This regulates the penetration of several substances by this enzyme. Hence, this could be

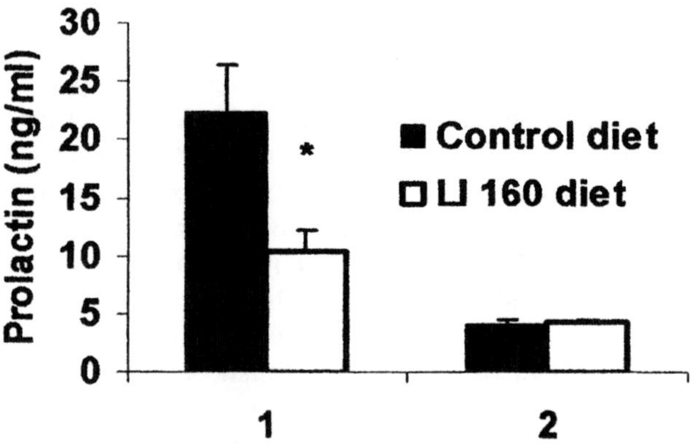

Figure 5. Effect of a diet containing LI 160 (3 g/kg) *versus* control diet on the serum prolactin responses to the 5-HT_{2A} agonist, DOI (dose = 2 mg/kg *ip*) and saline (*ip*) in male rats (n = 7 per group). Results are given as mean ± sem. * $p < 0.04$. Taken from [17].

another alternative process by which LI 160 may be reducing the neuroendocrine responses to DOI (i.e., by a direct reduction of the DOI content in the brain itself).

Long-term treatment with HP extract appears to reduce the hypothalamic–pituitary axis (HPA) responsiveness in terms of cortisol or ACTH release in animal studies [19–24]. For example, one very recent study reported the effects of Jarsin (LI 160) at two dose levels (600 mg and 1,800 mg per day) in healthy male volunteers on evening salivary cortisol and melatonin concentrations [25]. This study was carried out to test the hypothesis that HP extracts exert their antidepressant actions via DA and 5-HT-mediated pathways but not through NA as previous studies from the same laboratory have shown. In this study the subjects acted as their own controls. Overall, results showed that Jarsin treatment increased evening salivary cortisol; however this was only significant at the lower dose (see Figs 6a and b). The findings suggest that Jarsin increases evening salivary cortisol output in a U-shaped dose-dependent manner. There is some animal data in the literature to support this finding [26–28]. Neither of the doses of Jarsin used increased saliva melatonin concentrations. The authors maintained that the results from this study and previous work of theirs and others [10, 12, 17, 23, 29] upholds the notion that the HP extract LI 160 acts mainly through DA and 5-HT-mediated receptor pathways and did not support a role for NA in its mode of action. They also suggested that Jarsin may have antidepressant efficacy at a lower dose than the presently used daily recommended dose of 900 mg, which would potentially offset the already low side effect profile of such treatment.

In direct contrast to the previously discussed study in human volunteers, studies in animals show that longer-term treatment as stated earlier, tend to decrease HPA activity in terms of corticosterone and ACTH responsiveness. The most obvious reason for this difference may be that the period of treatment in the human volunteer study above could have been too short to bring about any possible receptor adaptation effects in terms of desensitisation or down-regulation. Whereas in animals studies reported, the treatment periods with HP extracts or its constituents have been for a minimum 2 weeks, and indeed in some instances up to 8 weeks. Another reason for the contrasting results between human and animals studies may simply be down to species differences alone.

One such study showed that treatment with the HP extract LI 160, hypericin and imipramine for 14 days in rats significantly reduced both ACTH and corticosterone concentrations when compared to placebo treatment (see Fig. 7) [19]; while in another study by the group, 8 weeks treatment with the same compounds in rats significantly decreased CRH mRNA in the hypothalamic paraventricular nucleus (PVN) and similarly 5-HT_{1A} mRNA in the hippocampus [22]. However, none of these compounds caused an effect on these parameters after 2 weeks treatment. In a more recent study, the same group demonstrated that neither hyperforin, an analogue of hyperforin or haloperidol (a negative control for the study) had any significant effects on CRF, glucocorti-

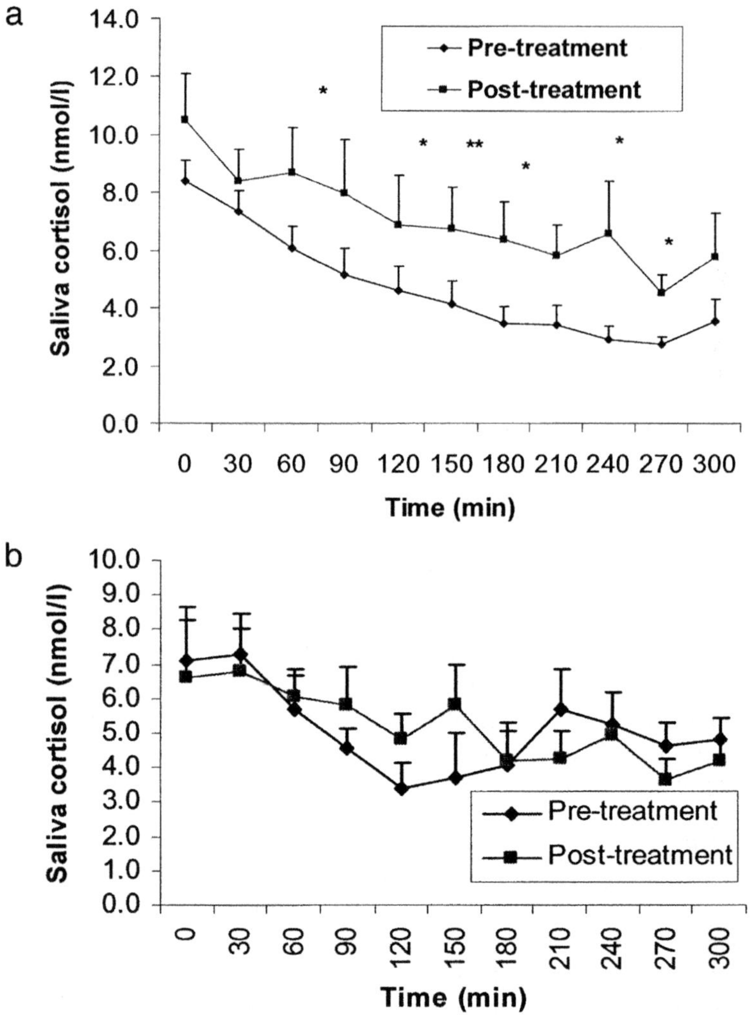

Figure 6 (a and b). Effect of low and high-dose treatment (600 and 1,800 mg/day respectively) with Jarsin for 1 week on evening salivary cortisol concentrations on the day before and on the last day of treatment. Results are mean ± sem and n = 9 per group. * p = 0.05, ** p = 0.01. Taken from [25].

coid, mineral-corticoid or 5-HT_{1A} gene expression [30], whereas the positive control in this study, fluoxetine, significantly increased all of these after 8 weeks treatment, but not after 2 weeks. The authors maintained that hyperforin and its derivatives are not involved in the regulation of genes that control HPA axis function.

Two further studies lend support to these findings. In the first, an extract of HP was shown to improve the resistance to stress in rats [24]. They showed that rats treated for 30 days with an HP extract had reduced ACTH and adre-

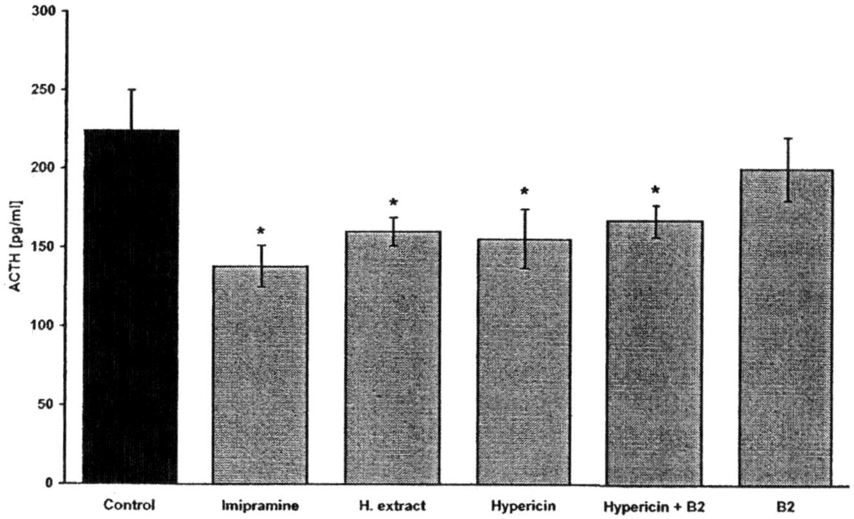

Figure 7. Plasma ACTH levels after 14 days of daily pretreatment (n = 8) per group; $*p < 0.05$ *versus* control). Taken from [22].

nal weight compared to animals treated with a placebo following stress treatment. In the second study subchronic (2 weeks) treatment with the HP extract LI 160 and imipramine significantly reduced serum corticosterone responses as compared to those for similar acute, single dose treatments [23]. In the case of LI 160, this was dose-dependent. The authors concluded that based on evidence from previous blockade studies of theirs [12], that this was most probably due to downregulation of postsynaptic 5-HT_2 receptors.

A recently published study has demonstrated that subchronic treatment with the HP extract LI 160 significantly reduces both corticosterone and cortisol in rat brain frontal cortex tissue [31]. It was suggested by the authors that this might be caused by HP's ability to increase the expression of the multidrug transporter glycoprotein (Pgp) at the level of the blood–brain barrier (BBB), an effect which has been demonstrated by others [32–34]. It was proposed that this might be another way in which HP could be expressing its mode of antidepressant action, since cortisol is often raised in plasma and cerebral spinal fluid (CSF) during depression [35] and normalises upon recovery [36].

Reductions in 5-HT_{2A} receptor sensitivity have been seen following chronic treatment with certain antidepressants [37–39]. However, few studies have described the effect of chronic treatment with newer conventional therapies such as the selective serotonin reuptake inhibitors (SSRIs) on neuroendocrine responses to 5-HT_2 agonists. The composite data suggests that long-term treatment with SSRIs tends to increase the sensitivity or maximal responsiveness of hormones such as corticosterone, ACTH, oxytocin and to a lesser degree prolactin after challenges with DOI or MK-212 [40]. One recent study on human volunteers looked at the effect of both short-term (7 days) and longer-term (21 days) treatment with paroxetine on the cortisol responses to the indirect 5-HT_2 agonist, 5-HTP [41]. In this study, short-term treatment enhanced plasma cortisol when compared to pretreatment. However, longer-term treatment reduced cortisol release compared to short-term treatment, although it was still greater than pretreatment. Hence, there may be some suggestion here of a delayed desensitisation of post-synaptic 5-HT_2 receptors as the term of treatment lengthens. However, it is largely held that present studies suggest a general potentiation of 5-HT_2 neurotransmission.

Prolactin

As previously discussed in the text, HP extracts inhibit the reuptake of DA, 5-HT and DA [5]. However, it was obvious from these same studies that HP more potently inhibited DA than either 5-HT or NA reuptake. Prolactin is released from the anterior pituitary gland and is under strong inhibitory control by prolactin inhibiting factor (PIF), thought possibly to be DA itself. It is under release control by 5-HT [1]. However, the inhibitory control on prolactin secretion is the greater of the two modulations.

In studies carried out in healthy human volunteers the HP extracts Jarsin (LI 160) and WS5570 were shown to reduce plasma prolactin concentrations [7, 10]. In the first study, a single dose of 2,700 mg of Jarsin (three times the recommended normal daily intake of 900 mg) was shown to significantly decrease serum prolactin concentrations (see Fig. 8) [10]. The authors suggested that the explanation for this was that Jarsin enhances some aspects of DA neurotransmission. This is because DA pathways facilitate the inhibition of prolactin secretion [1]. Since most conventional, antidepressants acutely do not increase DA neurotransmission as Jarsin obviously does, it was suggested by the authors that this might be of some significance. Of the currently available antidepressant drugs, only bupropion is believed to possess significant DA action, which is probably due to DA reuptake blockade [42–44]. Interestingly, the clinical profile of activity of bupropion is rather different to that of other antidepressants that facilitate NA and 5-HT. For example, bupropion may have lesser risk of producing rapid cycling in bipolar disorder [39] and also has efficacy in the management of smoking cessation [45]. It is therefore conceivable that HP extracts with a higher percentage of hyperforin such as LI 160 content might also be useful in the management of smoking cessation. Additionally, with its 5-HT and NA reuptake blocking action it may be of further benefit in helping to dampen possible changes in mood and anxiety state that usually occur during nicotine withdrawal.

In a second study carried out in healthy human male volunteers, the investigators failed to show significant dose-dependent decreases in plasma pro-

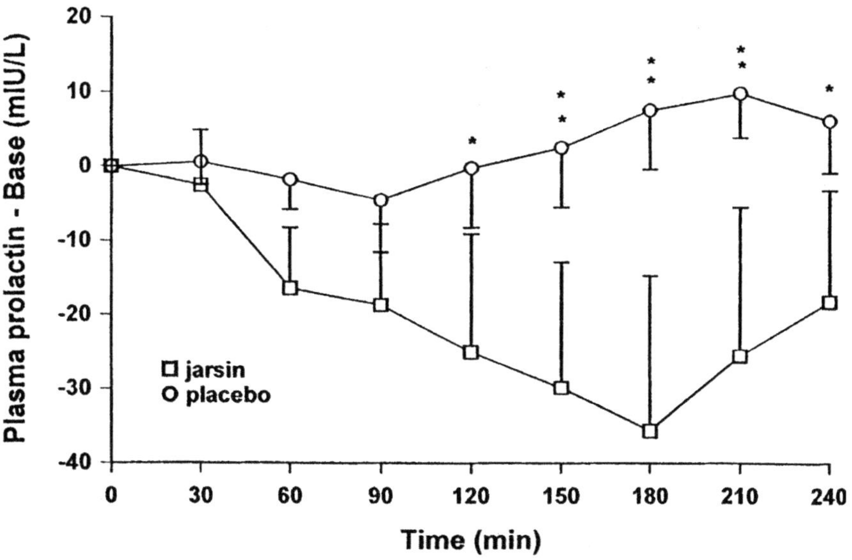

Figure 8. Effect of HP extract (LI 160, 2,700 mg) administered at time 0 and placebo on plasma prolactin (mean ± sem) measured as change from baseline over time in 12 healthy male volunteers. The ANOVA showed no main effects of time ($F = 0.9$; $df = 70.7$; $p = 0.42$) however there were significant effects of drug ($F = 3.7$; $df = 10.1$; $p = 0.045$) and drug by time interaction ($F = 2.1$; $df = 70.7$; $p = 0.035$). Significantly different from placebo * $p < 0.05$ and ** $p < 0.01$ (Fisher's test of least significant difference). Taken from [23].

lactin following acute challenge with doses of 300 mg and 600 mg, respectively, of WS5570 *versus* placebo [7]. However, there was a non-significant reduction in plasma prolactin when the area under the dose-response curve data was collated (Fig. 9). In a later study from the same laboratory, no significant effect of the extract WS5570 at various doses *versus* placebo could be found in healthy volunteer subjects [9]. The authors propose an interplay between the opposing inhibitory and stimulatory modulation effects of the HP extract on DA and 5-HT respectively for the negative findings on plasma prolactin excretion in this reported new study.

Overall, the results from these studies are probably not directly comparable to that previously discussed [10], since different HP extracts were used. However, taken together the studies do show that HP extracts reduce plasma prolactin. Additionally, in our study [10] the dose used was much higher than that used by Schule et al. [7] and therefore might be expected to produce a greater neuroendocrine response. Interestingly, there is evidence to suggest that HP extracts may produce U-shaped dose dependent effects [27, 28].

In an acute study in rats, both the HP extract LI 160 and hyperforin were found to potently reduce serum prolactin secretion, whilst hypericin was found to cause a non-significant increase [12] (Tab. 3). In the same study LI 160 and hyperforin pretreatment significantly reduced the normal potent prolactin

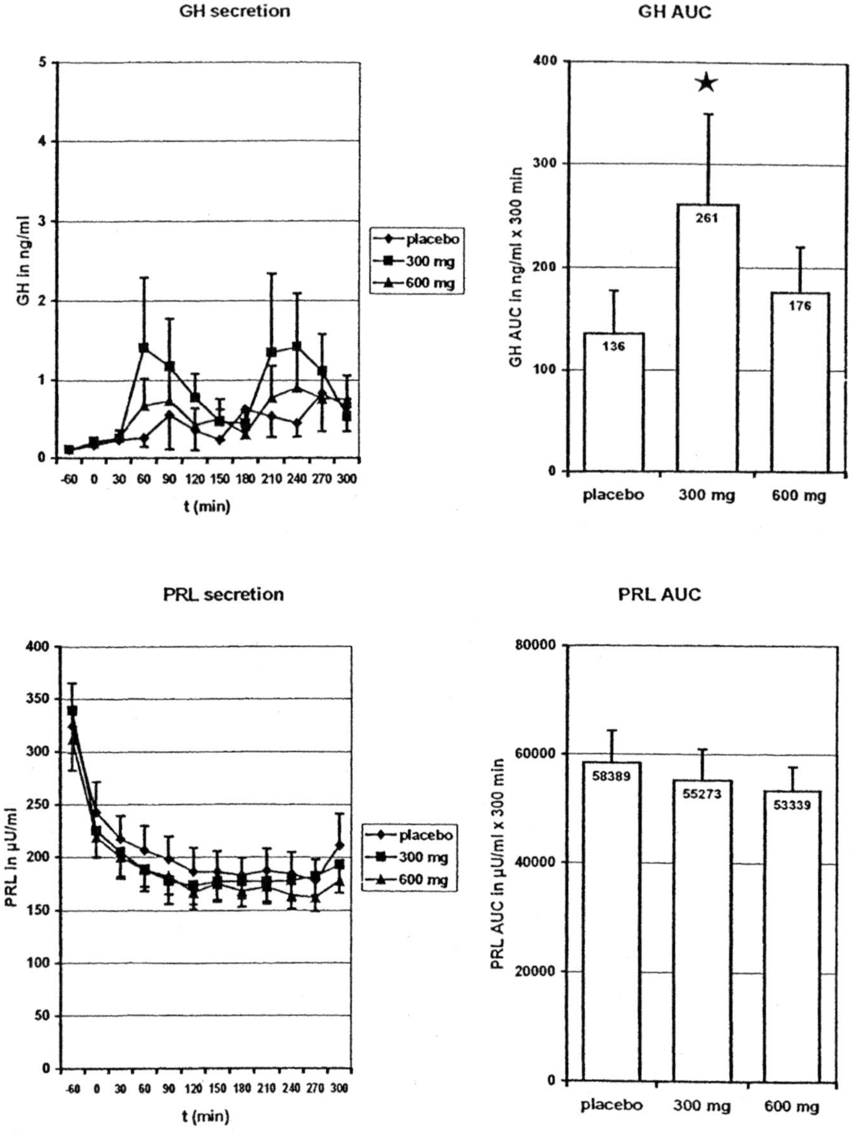

Figure 9. Growth hormone (ng/ml) prolactin (MU/ml) after administration of placebo, WS 5570 300 and 600 mg; mean value and AUC graphs. S.E. Indicated; star: p = 0.05 (*versus* placebo). Taken from [7].

responses to the DA antagonist and psychotropic drug, haloperidol (Fig. 10). Pretreatment with hypericin, (not shown in the figure) had no effect on the prolactin response to haloperidol. The authors proposed that the prolactin response to LI 160 and hyperforin were DA-mediated. The study by

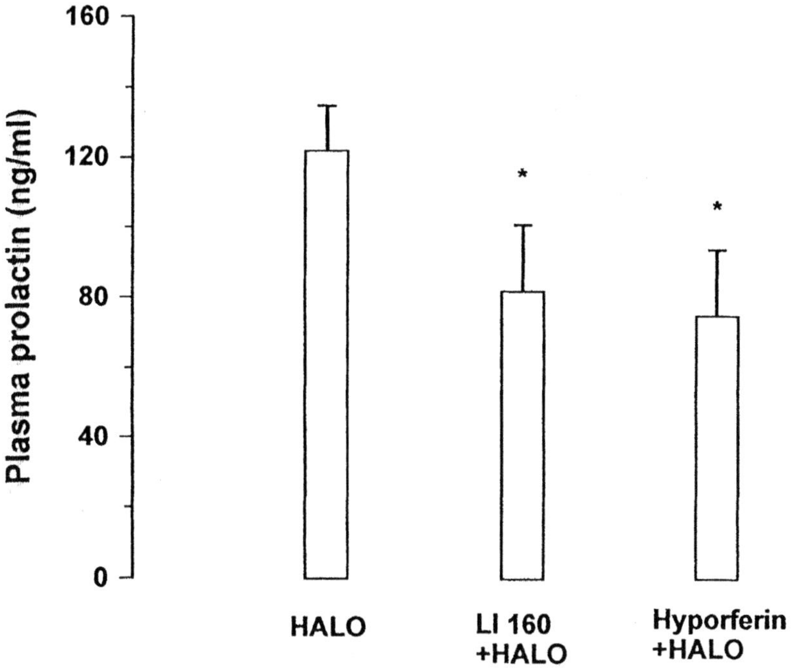

Figure 10. Effect of LI 160 (200 mg/kg, *ip*) on the plasma prolactin responses to the dopamine antagonist, haloperidol (0.5 mg/kg, *ip*), LI 160 and hyperforin were administered 30 min prior to treatment with haloperidol and the animals were sacrificed 3, 5 and 2.5 h respectively after this. * $p < 0.01$. Taken from [23].

Butterweck et al. [46] lends some support to these findings. They showed that LI 160 extract strongly inhibited thyroid releasing hormone (TRH) stimulated release of prolactin from primary pituitary cell cultures.

In the same paper the authors also showed that 2 weeks treatment with LI 160 by gavage caused significant reductions in plasma prolactin concentrations in rats. In a further study by the same authors, repeated treatment for 2 weeks with LI 160, hypericin and imipramine produced no alteration in plasma prolactin concentrations as compared to placebo [19].

Acutely, the effects of conventional antidepressant drugs in both animals and man on plasma prolactin secretion are highly variable and may or may not be related to their antidepressant mode of action [47–49]. The SSRI citalopram effects a dose-dependent increase in plasma prolactin following single intravenous administration of the drug in healthy human volunteers [48]. Somewhat surprisingly, the supposedly highly selective noradrenaline reuptake inhibitor (SNRI) reboxetine also stimulates the secretion of plasma prolactin [49], whereas the "atypical" antidepressant, mirtazapine shows no effect [47] and bupropion indeed reduces plasma prolactin [50].

Growth hormone

Growth hormone (GH) like prolactin is under dual release control. Growth hormone releasing hormone (GHRH) and somatostatin are secreted from the hypothalamus and modulate the pulsatile release of GH from the anterior pituitary by stimulatory and inhibitory mechanisms, respectively. GHRH is probably influenced by both NA and 5-HT activation, while somatostatin is mediated via DA.

Studies in healthy human volunteers suggest that HP extracts stimulate the release of GH but that responses may be variable and not necessarily robust [7, 10]. In the study by Schule and colleagues [7] subjects were treated with doses of 300 mg and 600 mg of the HP extract WS5570 and placebo on separate occasions. Results showed that the 300 mg dose caused a significant increase in plasma GH output according to area under the plasma GH profile curve (AUC) but that the 600 mg dose rather surprisingly caused no change and was similar to placebo (Fig. 9). In a second and later study in healthy volunteers, the same authors were unable to completely reproduce their previous findings in that the extract WS5570 did not cause significant stimulation of serum GH secretion [9]. However, results from the study showed that the extract WS5570 did cause non-significant but obvious late increases (in terms of time from administration) in serum GH, but these were not dose-related. Overall, the data from these studies may suggest that this HP extract stimulates the release of plasma GH in a nonlinear and possibly U-shaped dose-dependent manner similar to cortisol and prolactin. Similar findings have been reported before [25, 27, 28]. In the only other reported study concerning HP and GH release, it was shown that a single high dose (2,700 mg) administration of the HP extract Jarsin in healthy male volunteers caused a significant increase in plasma GH concentrations as compared to placebo which peaked approximately 2 h after dosing (Fig. 11) [7]. Analysis of plasma concentrations of both hypericin and hyperforin over the same time period of the study suggested that the GH responses may be related to hyperforin plasma concentration kinetic change but not to that of hypericin. Hence it is possible that HP's ability to stimulate GH release may be explained by the DA reuptake properties of the constituent, hyperforin, as DA pathways are known to facilitate GH release [1].

Other hormones

Melatonin, the hormone of the pineal gland, is controlled by the release of NA from pre-junctional sympathetic terminals which synapse with beta-adrenoceptors in pinealocytes. In humans, melatonin is secreted during the hours of darkness, peaking at around 02.00 h. The major metabolite of melatonin, 6-sulphatoxymelatonin, which is excreted in urine, is a reliable and valid measure of melatonin secretion [29].

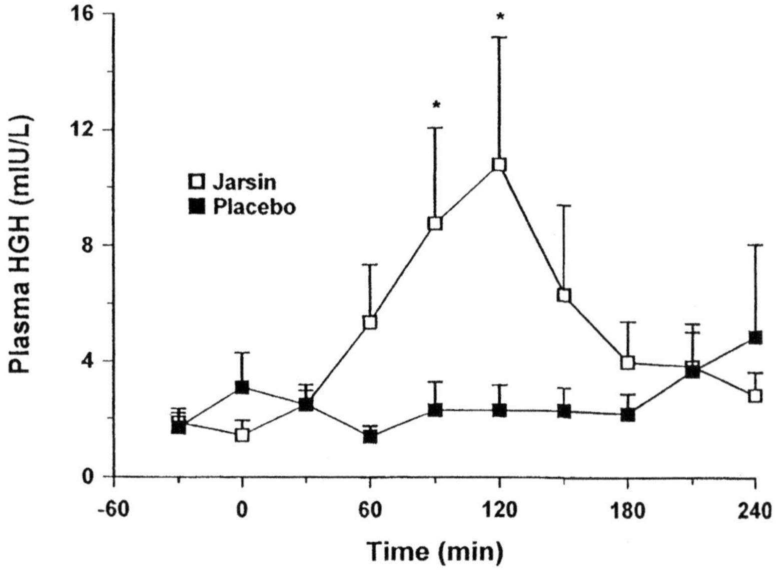

Figure 11. Effect of HP extract (LI 160, 2,700 mg) administered at time 0 and placebo on plasma HGH (mean ± sem) measured as change from baseline over time in 12 healthy male volunteers. The ANOVA showed significant main effects of drug ($F = 5.6$; $df = 7.1$; $p = 0.05$) and time ($F = 2.2$; $df = 63.9$; $p = 0.036$) and there was also a significant effect of drug by time interaction ($F = 2.3$; $df = 63.9$; $p = 0.024$). Significantly different from placebo ** $p < 0.01$ (Fishers test of least significant difference). Taken from [23].

One recent study looked at the effects of two doses of the HP extract Jarsin (LI 160) in healthy male volunteers on evening salivary levels of cortisol and melatonin and 6-sulphatoxymelatonin in urine. Results showed that neither the high (1,800 mg) nor the low dose (600 mg) had any significant effect on either of the melatonin measures. In the only reported animal study the effects of various doses (0–200 mg/kg rat body weight) of the HP extract LI 160 and the NA reuptake inhibitor desipramine (DMI, 10 mg/kg) on the pineal content of melatonin were determined [23]. Results showed that LI 160 had no significant effect on pineal melatonin at any dose; whereas the positive control drug DMI caused a four-fold increase (Tab. 4), a result which was similar to previous studies for DMI [51]. The studies taken together suggest that the HP extract tested does not effect melatonin secretion, even at higher doses. Therefore it is unlikely that HP's actions are mediated via NA receptor pathways.

Two weeks treatment by gavage with the HP extract LI 160, hypericin or imipramine failed to alter plasma luteinising hormone (LH) concentrations in rats [22]. In the same animals testes and seminal vesicle weights were also unaffected by any of the repeated treatments and suggests, according to the authors, that testosterone synthesis is generally not affected either.

Table 4. Effect of various doses of LI 160 extract and DMI (10 mg/kg) on pineal melatonin content in the rat

Dose mg/kg					
0	20	50	100	200	DMI (10)
178 ± 21	143 ± 42	165 ± 36	101 ± 20	143 ± 27	*806 ± 109

Each point represents the mean ± sem pg/pineal melatonin. Analysis of the data by ANOVA showed no significant effect of dose ($F = 1.11$; $df = 4, 18$; $p = 0.36$). * $p < 0.001$. Taken from [22]

Another study report suggests that HP may elevate thyroid stimulating hormone (TSH) in humans [52]. However, due to the obscurity of the study, the results which were obtained via a standardised telephone interview to 37 subjects should perhaps be taken with some degree of scepticism.

Summary

Overall, acute treatment studies with various extracts of HP suggest that its antidepressant effects are probably mediated in the main through alterations in 5-HT and DA neurotransmission and do not generally involve NA. This is certainly the case in the time frame in which the studies reported here have been carried out in. With respect to DA-mediated processes, studies have demonstrated that HP extracts decrease plasma prolactin in both animals and man and that this is dose-dependent in the rat. It also increases GH in man. Of the active ingredients in HP extracts studied, hyperforin but not hypericin was shown to decrease plasma prolactin; biochemical, pharmacokinetic and clinical findings lend support to these findings. The HP extract LI 160, hyperforin and hypericin all increase serum corticosterone in the rat, which from blockade studies appeared to be mediated via a post-synaptic 5-HT_2 receptor mechanism. LI 160 also increased salivary cortisol in man. Some hormone responses to HP extract single dose and subchronic administration demonstrated nonlinear dose-response relationships and may possibly be U-shaped in nature.

Sub-chronic treatment studies in the rat showed reduced serum corticosterone and increased serum prolactin responses when compared to acute treatment with an HP extract at two different doses. Further, both serum corticosterone and prolactin responses to neuroendocrine challenge with the 5-HT_{2A} agonist DOI following treatment for 2 weeks with an extract of HP, were reduced compared to placebo. Thus post-synaptic 5-HT_2 receptors may downregulate or are desensitised with respect to corticosterone output and that DA inhibitory control of prolactin release may be reduced due to increases in 5-HT-mediated influences. CRF mRNA gene expression studies lend support to these findings. In a recent clinical trial carried out in healthy male volunteers, salivary cortisol response-profiles following 7 days treatment with two different doses of the extract LI 160 were increased, but this was greater at the lower dose. The authors suggested that HP may exhibit a nonlinear dose-response U-shaped relationship with respect to salivary cortisol output. In the same study the rise in evening salivary melatonin was not altered by either dose. It was suggested that since melatonin output is mostly NA driven, that treatment with HP, certainly in the shorter term, probably does not effect NA neurotransmission.

Further studies are required to confirm 5-HT and DA-mediated mechanisms of action of HP extracts utilising the neuroendocrine paradigm and to access more fully dose-response relationships. Studies have shown that hyperforin is probably the most important active ingredient of HP extracts. It is therefore perhaps important that clinical trials are performed in patients with mild-to-moderate depression to show that the efficacy of the HP extract under test is dependent on its hyperforin content, so that future HP treatments may be standardised on hyperforin content and not the present hypericin content.

References

1 Tuomisto J, Mannisto P (1985) Neurotransmitter regulation of anterior pituitary hormones. *Pharmacol Rev* 37: 249–332
2 Checkley SA (1980) Neuroendocrine tests of monoamine function in man: a review of basic theory and its application to the study of depressive illness. *Psychol Med* 10: 35–53
3 Yatham LN, Steiner M (1993) Neuroendocrine probes of serotonergic function: a critical review. *Life Sci* 53: 447–463
4 Vorbach EU, Arnoldt KH, Hubner WD (1997) Efficacy and tolerability of St John's Wort extract LI 160 *versus* imipramine in patients with severe depressive episodes according to ICD-10. *Pharmacopsychiatry* 30 (Suppl 2): 81–85
5 Müller WE, Rolli M, Schafer C, Hafner U (1997) Effects of *Hypericum* extract (LI 160) in biochemical models of antidepressant activity. *Pharmacopsychiatry* 30 (Suppl 2): 102–107
6 Franklin M, McGavin C, Reed A, Cowen PJ (1998) Effect of *Hypericum perforatum* on salivary cortisol in healthy male volunteers. *J Psychopharmacol* 30 (Suppl): A16
7 Schule C, Baghai T, Ferrera A, Laakmann G (2001) Neuroendocrine effects of *Hypericum* extract WS 5570 in 12 healthy volunteers. *Pharmacopsychiatry* 34 (Suppl): 127–133
8 Laakmann G (1987) *Psychopharmacoendocrinology and Depression Research*. Springer, Berlin, Heidelberg, New York
9 Schule C, Baghai T, Sauer N, Laakmann G (2004) Endocrinological effects of high-dose *Hypericum perforatum* extract WS5570 in healthy subjects. *Neuropsychobiology* 49: 58–63
10 Franklin M, Chi J, McGavin C, Hockney R, Reed A, Campling G, Whale RWR, Cowen PJ (1999) Neuroendocrine evidence for dopaminergic actions of *Hypericum* extract (LI 160) in healthy volunteers. *Biol Psychiatry* 46: 581–584
11 Müller WE, Singer A, Wonnermann M, Hafner U, Rolli M, Schafer C (1998) Hyperforin represents the neurotransmitter re-uptake inhibiting constituent of *Hypericum* extract. *Pharmacopsychiatry* 31 (Suppl 1): 16–21
12 Franklin M, Chi J, Mannel M, Cowen PJ (2000) Acute effects of LI 160 (extract of *Hypericum perforatum*, St John's Wort) and two of its constituents on neuroendocrine responses in the rat. *J Psychopharmacol* 14: 360–363
13 Calapai G, Crupi A, Firenzuoli F, Costantino G, Inferrera G, Campo GM, Caputi AP (1999) Effects of *Hypericum perforatum* on levels of 5-hydroxytryptamine, noradrenaline, dopamine in the cortex, diencephalons and brain stem of rats. *J Pharm Pharmacol* 51: 723–728
14 Kaehler ST, Sinner C, Chattergee SS, Philippu A (1999) Hyperforin enhances serotonin and glu-

tamate in the rat locus ceoruleus. *Neurosci Lett* 262: 199–202
15 Di Matteo V, Di Giovanni G, Di Mascio M, Esposito E (2000) Effect of acute administration of hypericum-CO_2 extract on dopamine and serotonin release in the rat central nervous system. *Pharmacopsychiatry* 33: 14–18
16 Kehr J, Yoshitake T, Yoshitake S, Ogren SO (2003) Differential effects of acute and repeated administration of *Hypericum perforatum* (St John's Wort) on extracellular levels of serotonin, noradrenaline and dopamine in pre-frontal cortex, striatum and hippocampus in the rat. *Eur Neuropsychopharmacol* 12: S203
17 Franklin M (2003) Sub-chronic treatment effects of an extract of *Hypericum perforatum* (St John's Wort, LI 160) on neuroendocrine responses to the 5-HT_{2A} agonist, DOI in the rat. *Pharmacopsychiatry* 36: 161–164
18 Bagdy G, Macara GB (1994) Hypothalamic paraventricular nucleus lesions differentially effect 5-HT_{1A} and 5-HT_2 receptor agonist-induced oxytocin, prolactin and corticosterone responses. *Endocrinology* 134: 1127–1131
19 Butterweck V, Winterhoff H, Herkanham M (2001) St John's Wort, hypericin and imipramine: a comparative analysis of mRNA levels in brain areas involved in HPA axis control following short-term and long-term administration in normal and stressed rats. *Mol Psychiatry* 6: 547–564
20 Hennessy M, Kellerher D. Spiers JP, Kavanagh P, Back D, Mulcahy F, Feely J (2002) St Johns Wort increases expression of P-glycoprotein: implications for drug interactions. *Br J Clin Pharmacol* 53: 75–82
21 Perloff MD, von Moltke LL, Stormer E, Shader RI, Greenblatt DJ (2001) Saint John's Wort: an *in vitro* analysis of P-glycoprotein induction due to extended exposure. *Br J Pharmacol* 134: 1601–1608
22 Butterweck V, Korte B, Winterhoff H (2001) Pharmacological and endocrine effect of *Hypericum perforatum* and hypericin after repeated treatment. *Pharmacopsychiatry* 34 (Suppl 1): 52–57
23 Franklin M, Cowen PJ (2001) Researching the antidepressant actions of *Hypericum perforatum* (St John's Wort) in animals and man. *Pharmacopsychiatry* 34 (Suppl 1): S29–S37
24 Makina DM, Taranukhin AG, Chernigovskaya EV, Kuzik VV (2001) Effect of *Hypericum* extract on the hypothalamic-pituitary adrenal system in rats. *Bull Exp Biol Med* 132: 1180–1181
25 Franklin M, Hafizi S, Reed A, Murck H (2004) Effect of sub-chronic treatment with Jarsin (extract of St John's Wort, *Hypericum perforatum*) at two dose levels on evening salivary melatonin and cortisol concentrations in healthy male volunteers. *Pharmacopsychiatry*, *in press*
26 Butterweck V, Wall A, Lieflander-Wulf U, Winrehohh H, Nahrstedt A (1997) Effects of the total extract and fractions of *Hypericum perforatum* in animal assays for antidepressant activity. *Pharamacopsychiatry* 30 (Suppl): 117–124
27 Rommelspacher H, Siemanowitz B, Mannel M (2001) Acute and chronic actions of a dry methanolic extract of *Hypericum perforatum* and a hyperforin-rich extract on dopaminergic and serotonergic neurones in rat nucleus accumbens. *Pharmacopsychiatry* 34 (Suppl 1): S119–126
28 Dimpfel W, Schober F, Mannel M (1998) Effects of a methanolic extract and a hyperforin-enriched extract of St John's Wort (*Hypericum perforatum*) on intracerebral field potentials in the freely moving rat (Tele-stereo-EEG). *Pharmacopsychiatry* 31 (Suppl 1): 30–35
29 Arendt J (1985) Mammalian pineal rhythms. *Pineal Res Rev* 3: 161–213
30 Butterweck V, Winterhoff H, Herkenham M (2003) Hyperforin-containing extracts of St John's Wort fail to alter gene transcription in brain areas involved in HPA axis control: a long-term regimen in rats. *Neuropsychopharmacology* 28: 2160–2168
31 Franklin M, Reed A, Murck H (2004) Sub-chronic treatment with an extract of *Hypericum perforatum* (St John's Wort) significantly reduces cortisol and corticosterone in rat brain. *Eur Neuropsychopharmacol* 14: 7–10
32 Durr D, Steiger B, Kullak-Ublick BA, Rentsch KM, Steinert HC, Meier PJ, Fattinger K (2000) St John's Wort induces intestinal P-glycoprotein and intestinal and hepatic CYP3A4. *Clin Pharmacol Ther* 68: 598–604
33 Perloff MD, von Moltke LL, Stormer E, Shader RI, Greenblatt DJ (2001) St John's Wort: an *in vitro* analysis of p-glycoprotein induction due to extended exposure. *Bt J Pharmacol* 134: 1601–1606
34 Hennessy M, Kellerher D, Spiers JP, Kavanagh P, Back D, Mulcahy F, Feely J (2002) St John's Wort increases expression of p-glycoprotein: implications for drug interactions. *Brit J Clin Pharmacol* 53: 75–82
35 Carroll BJ, Curtis GC, Mendels J (1976) Cerebrospinal fluid and plasma free cortisol concentra-

tions in depression. *Psychol Med* 6: 235–244
36 Steiger A, Von Bardeleben U, Wiedemann K, Holsboer F (1991) Sleep EEG and nocturnal secretion of testosterone and cortisol in patients with major depression during acute phase and remission. *J Psychiatr Res* 25: 169–177
37 Meltzer HY (1984) Serotonergic function in the affective disorders: the effect of antidepressants and lithium on the 5-HTP-induced increase in serum cortisol. *Ann NY Acad Sci* 430: 115–137
38 Gartside SE, Ellis PM, Sharp T, Cowen PJ (1992) Selective 5-HT_{1A} and 5-HT_2 receptor-mediated adrenocorticotropin release in the rat: effect of repeated antidepressant treatments. *Eur J Pharmacol* 221: 103–109
39 Maj J, Bijak M, Dziedzicka-Waslewska M, Rogoz R, Rogoz Z, Skuza G, Tokarski T (1996) The effects of paroxetine given repeatedly on the 5-HT receptor subpopulations in the rat. *Psychopharmacology* 127: 73–82
40 Raap DK, Van de Kar LD (1999) Selective serotonin re-uptake inhibitors and neuroendocrine function. *Life Sci* 65: 1217–1235
41 Sargent PA, Williamson DJ, Cowen PJ (1998) brain 5-HT neurotransmission during paroxetine treatment. *Brit J Psychiatry* 172: 49–52
42 Arnett CD, Wolf AP, Shine H (1986) Improved delineation of human dopamine receptors using ^{18}F-N-methylspiroperidol and PET. *J Nucl Med* 27: 1878–1886
43 Ascher JA, Cole JA, Colin JN, Feighner JP, Ferris RM, Fiberger HC, Golden RN, Martin P, Potter WL, Richelson E, Sulser F (1995) A review of its mechanism of antidepressant activity. *J Clin Psychiatry* 56: 395–401
44 Potter WL (1998) Bipolar depression: specific treatments. *J Clin Psychiatry* 59: 30–38
45 Hurt RD, Sachs DPL, Glover ED, Offord KP, Johnson JA, Loweell CD (1997) A comparison of sustained-release bupropion and placebo for smoking cessation. *N Engl J Med* 337: 1195–1202
46 Butterweck V, Nahrstedt A, Winterhoff H (1998) Effects of *Hypericum perforatum* extract and isolated constituents on neurotransmitter concentrations and endocrine parameters. *Ann Congress of the Soc of Med Plant Res*: J29
47 Laakmann G, Schule C, Baghai T, Waldvogel E (1999) Effects of mirtazapine on growth hormone, prolactin and cortisol secretion in healthy male subjects. *Psychoneuroendocrinology* 24: 769–784
48 Attenburrow MJ, Mitter PR, Whale R, Terao T, Cowen PJ (2001) Low-dose citalopram as a 5-HT neuroendocrine probe. *Psychopharmacology* 155: 323–326
49 Schule C, Baghai T, Schmidbauer S, Bidlingmaier M, Strasburger C. Laakmann G (2004) Reboxetine acutely stimulates cortisol, ACTH, growth hormone and prolactin secretion in healthy male subjects. *Psychoneuroendocrinology* 29: 185–200
50 Melzer HY, Simonovic M, Sturgeon RD, Fang VS (1981) Effect of antidepressants, lithium and electroconvulsive treatment on rat serum prolactin. *Acta Psychiatr Scand Suppl* 290: 100–121
51 Burns TG, Brown GM (1985) The effects of acute and chronic desipramine treatment on pineal and serum melatonin and N-acetylserotonin. In: GM Brown, SD Wainwright (eds): *The Pineal gland*. Pergamon Press, Oxford, UK, 25–30.
52 Ferko N, Levine MA (2001) Evaluation of the association between St John's Wort and elevated thyroid-stimulating hormone. *Pharmacotherapy* 21: 1574–1578

Index

GPSR Compliance

The European Union's (EU) General Product Safety Regulation (GPSR) is a set of rules that requires consumer products to be safe and our obligations to ensure this.

If you have any concerns about our products, you can contact us on ProductSafety@springernature.com

In case Publisher is established outside the EU, the EU authorized representative is:

Springer Nature Customer Service Center GmbH
Europaplatz 3
69115 Heidelberg, Germany

Batch number: 10732314

Printed by Printforce, the Netherlands